T5-AXO-731

MANAGING INNOVATION

Managing Innovation

From the Executive Suite to the Shop Floor

John S. Rydz

Ballinger Publishing Company • Cambridge, Massachusetts
A Subsidiary of Harper & Row, Publishers, Inc.

Copyright © 1986 by Ballinger Publishing Company. All rights reserved. No part of this publication may be reproduced, stored in a retrieval system, or transmitted in any form or by any means, electronic, mechanical, photocopy, recording or otherwise, without the prior written consent of the publisher.

International Standard Book Number. 0-88730-028-6

Library of Congress Catalog Card Number: 85-1156

Printed in the United States of America

Library of Congress Cataloging-in-Publication Data
Rydz, John S.
Managing innovation.
Includes index.
1. Technological innovations—Management. I. Title. HD45.R93
1986 658.4'06 86-1156
ISBN 0-88730-028-6

To Brian and David

Two future innovators

CONTENTS

LIST OF FIGURES

INTRODUCTION

Daniel Boorstin, the Pulitzer Prize-winning author and historian, has remarked that "the most promising words ever written on the maps of human knowledge are *terra incognita*—unknown territory." If that is the case, then those of us involved in the world of American business are facing the most promising time in history because we have never faced as much unknown territory as we do today.

For the first time, change is the rule rather than the exception. Thanks to the phenomenon of constant change at the global level, we in business and industry are seeing extraordinary changes every day—changes that will repeal the present rules of the game, reorient the direction of the business, restructure the work force, redefine the meaning of work itself, redraw organizational charts, and recast the participants into new roles.

As a result, the office desk, the factory floor, the engineering drawing board, the customer sales order form, and thousands of other details that make up commerce in the modern world will never be the same again. Each of these classic business tools has already changed so much that managers of only a decade ago would not recognize them. As these changes take place, new organizational configurations are replacing old-fashioned formulas and dogmas. Indeed, as I have discussed the effects of change with other business executives, many have told me that half *or more* of the jobs in their companies will change in the next five years.

These changes are challenging managers and executives as never before and, sadly, many of them do not appear to be up to the task. People have a natural resistance to change anyway, and, to those who have come to think of the world as a stable place, change can be disorienting. When I think back to management conferences with colleagues of mine at The Singer Company—conferences dedicated to the understanding of change—I realize that over half of the vice presidents attending are no longer with that company. This situation gives first-hand example of the impact of change.

Clearly, however, change cannot be avoided. We can no longer rhetorically point to the coming storm and call for acceptance of the new order of things at some point in the future. The storm is upon us—and it has been upon some companies for thirty years or more.

I, for one, do not believe that change is something to be afraid of. Like Daniel Boorstin, I believe *terra incognita* always holds promise. In my career with Emhart, RCA, Singer, Diebold, and other companies, my job has been to confront change directly—to stimulate, manage, and harness it to create new products and new profits. And I believe that, more than ever, managers who must confront change can do so successfully and, just as important, have fun while they're doing it. They can do so if they understand that change is the raw material of innovation, and that as a result innovation is moving to center stage in the business world. Innovation—staying ahead of the competition with better management, technology, and marketing, based on scheduled, managed creativity and follow-through—will be the hallmark of every company that hopes to survive and flourish to the end of the century.

And so this book is about managing change and innovation. It is about how some companies have combined these two elements to create continuous, even enviable success in business. It is also about how you can apply the lessons of those experiences to your company.

I want to convey these messages not only because I have enjoyed thirty years of successful management as a senior executive in some of the world's largest and best-known companies, but because I believe that in the future, managing is going to be even more exciting than in the past; the managers will have a greater role in influencing their companies' futures.

Another part of the message is that managers who rise to the challenge will be able to have a lot of fun along the way. I have always found that management is never dull—if it is approached as part of a process of innovation within the company. Very frankly, if managers are not enjoying their role, then there is something wrong, either with them or with the company.

Much of this book is anecdotal, telling the stories of companies that successfully innovated and survived—of managers who rose to the challenge of change—and of companies and managers that did not. In some companies, one corridor is full of success and jubilation at innovation, while down the hall the ship is sinking. So if I return to certain companies many times, it will be to look at different segments of each, to learn what worked in one area and what failed in another. We will cut apart several companies through these contraposed experiences to see how the pieces of the puzzle of success fit together.

Many of the stories I will tell involve companies that

innovated by converting from mechanical products to electronic products. But this does not mean that all problems in business can be solved by eliminating mechanical products and replacing them with electronic ones. Indeed, one of the most basic lessons in this book is that no manager—no company—should marry a particular type of technology, no matter what it is. The mechanical-to-electronic examples are plentiful because I was trained in electronics and also because I worked in the business world at a time when electronics was making many mechanical products obsolete. The underlying lessons would apply no matter what were the technologies involved.

Furthermore, there is nothing so magical about managing innovation that limits it to high-tech firms. The principles I am going to present to you are, very often, common-sense; they can be followed just as easily in basic manufacturing. In fact, one of the companies I will talk about—Nucor—successfully managed change by going *out* of high-tech businesses and innovating its way into steel production.

We will begin our journey by examining the nature of change—the raw material of innovation. In Part I, "The Change Imperative," we will examine the burgeoning impact of change and introduce the concept of the third derivative—the accelerating nature of change. We will see how one company, Diebold, manufacturer of automated teller machines, has learned to make change part of its corporate life. We will examine Singer's success with the electronic sewing machine as a means of understanding the concept of "quantum leap" innovation, and we will trace the success of RCA in developing color television in order to understand the concept of managing innovation.

In Part II, "Know Your Business," we will look at how change has altered the business world. We will discover that any business's customers, its competition, and even its own employees may no longer fit the traditional definitions. We will see how companies have failed in the last few years because they stuck to traditional definitions as change was swirling around them. Then we will examine how you must alter your own way of looking at things—your corporate culture—to understand what is happening.

In Part 3, "Innovate Your Business," we will learn the nuts and bolts of how to innovate change. We will introduce the concept of 3-D management and the technique of managing innovation by catalyzing people rather than managing them. Part 3 will also give you a short encyclopedia of techniques to introduce in your company to help stimulate innovation and excellence. Drawn from my experience and that of the companies described herein, they will enable you to remove the barriers to cultivating change and catalyzing people.

Welcome to this exciting world. Good luck. And remember to have fun.

PART I

The Change Imperative

1

THE CONSTANT OF CHANGE

Few American companies have as great a reputation for quality and as great a sense of mission as Diebold Incorporated. The company first earned this reputation during the great Chicago fire of 1871, when Diebold vaults and safes throughout the city protected their contents from the blaze and earned the company national recognition virtually overnight. Soon Diebold overshadowed all its competitors and, in the following decades, maintained its leadership position by dedicating itself to high-quality products and by carefully discerning and meeting the needs of the banking industry, no matter what those needs were.

I was aware of Diebold's reputation, but until I joined the firm as a vice president in 1965, I didn't understand how important this commitment to quality and meeting customer needs was to the company's success. My superior at Diebold was Raymond Koontz, the president, who, in his dark pinstripe suits, looked himself very much like a banker. A tough taskmaster, Koontz steered Diebold along a single-minded course, dedicating the company to supplying the finest equipment to the banking and finance industry. Koontz never lost sight of this single goal, and he never let his management team forget it. As a result, Diebold's managers fought the competition as if they were in a war. Mosler was their main competitor, and "Beat Mosler" was the essence of the Diebold spirit.

When I joined Diebold, Mosler had embarrassed the company by installing the vault doors at Fort Knox—the nation's pre-eminent symbol of bank security. My first assignment was to direct the research and development effort that would counter this coup: design and construction of the world's largest vault doors, the ones for the new headquarters building of the First National Bank of Chicago.

Those steel vault doors epitomized the Diebold corporate culture of 1965: strong, heavy, and mechanical. Yet, in a remarkably short period of time after that, Diebold changed its corporate culture completely in order to get a head start on the electronic revolution in banking. Despite a century of commitment to steel vaults, in only a few years Diebold became the leader in closed circuit television drive-in banking, electronic burglar alarms, telephone line security connecting bank alarms to police stations, and video recording surveillance systems. To this day, the company remains the market leader in automated teller machines (ATMs).

How did Diebold do it? How did the company move from a corporate culture that gloried in multi-ton vault doors to one that thrived on high-tech electronics and even more exotic sciences?

The answer is that Ray Koontz and his management team planned it that way. They created a culture within the company of *meeting needs.* By culture I mean the background of symbols, traditions, and myths that permeates any organization, and particularly a corporation. After a century of building vaults, Diebold's corporate culture might easily have been confined to the mechanics of steel. Yet the Diebold management team knew that meeting the needs of their customers was more important than their attachment to a particular type of technology. All of Diebold's managers were constantly on top of their customers' comments and requests. Any problem in the field with Diebold equipment became everyone's problem. Any product that was introduced was scrutinized by every company manager, simply because every manager knew that problems occurring with that product would ultimately become everyone's responsibility. Every idea for a new product or service was carefully and positively considered by every manager because they knew it could improve their relationship with Diebold's customers.

Marketing and sales managers knew that products sold to their customers had to perform. A product sold without adequate field testing or sufficient debugging brought down the wrath of all managers. Upon discovering problems with products, crash programs in engineering, manufacturing, and field service were instituted.

This insistence on reliable field performance made Diebold's R & D managers conscious of the need to have the best technology to meet these requirements. If mechanical technologies could not meet the demands of the banks, then new technologies were developed from whatever technological field worked best. There was no culture of any one technology. There was only a culture of meeting customer needs.

For example, the requirement to have secure telephone lines connecting branch bank alarms to police stations could be met only by using electronic equipment. A burglar's attack on a branch bank usually was preceded by cutting telephone lines, hence isolating the bank's alarm system from the police. By applying complex electrical or electronic signals on these telephone lines, police were immediately signaled when lines were cut. Diebold's competitors stayed with traditional techniques for providing this line security, but such equipment became vulnerable to sophisticated criminals. Complex, high-tech electronic coding proved to be far superior, and that is the direction Diebold took.

All managers within Diebold also were encouraged to participate fully in areas that acquainted them with the domains of critical change that affected their business. Strict attention was paid to state banking laws, which were moving in the 1960s toward liberalization of branch banking throughout the United States.

Additionally, Diebold's managers participated in conferences, seminars, and forums where society's trends and their impact on banking were analyzed. Senior management personnel met with key bankers visiting Diebold headquarters in Canton, Ohio; the objective was to learn what was happening in the banking industry. Any experiment being conducted at a bank with a competitive piece of technical equipment caught the attention of a sales manager at Diebold, who alerted other managers.

Due to this high degree of diligence, Diebold was able to recognize early the impact of smoke detectors, television drive-in banking, video recording for surveillance, sophisticated coding systems for security, and the banking needs for ATMs long before the competition.

R & D managers also were knowledgeable about banking operational requirements because they frequently toured banks. At one point, Earl Wearstler, the vice president of the bank division, saw that banks were concerned with the high turnover rates of tellers at the same time they were expanding their teller services. He relayed his observations to Diebold's R & D organization, which could use them in designing labor-saving equipment for banks.

The Diebold service organization became a good source of information about trends in operations in branch banks. In Washington, D.C., the people who serviced and sold Diebold equipment to intelligence networks also were responsible for keeping their ears to the ground to learn what was taking place in potential banking regulations.

During the early development stages of ATMs, Diebold kept in touch with the regulatory debate in Washington and was able to predict which regulations would apply to future ATM units. The same held true for contacts throughout all of the state legislatures. The field managers were constantly alert to any changes in bank laws, regulations, and the influence of the American Bankers Association. The network of marketing, sales, and services managers tracked bank expansion plans and excursions into new technology. Harry Parr, Diebold's financial vice president, also developed his network throughout his banking contacts, illustrating how all executives worked as a team, alerting everyone to bits and pieces of information that meant potential business for Diebold.

Close contacts were established with the Federal Bureau of Investigation and other government security agencies, thus providing input about the security of Diebold's equipment and the competition. Programs were established with leading research and development organizations, such as Battelle Institute, for the development of products that would resist torch and drill attacks, as well as new electronic equipment offering banks more efficient ways to conduct business. Initial investigations by research organizations led to early prototypes for ATMs. And joint ventures with electronic companies led to the development of sophisticated electronics equipment, ultimately marketed by Diebold.

Diebold has also expanded in areas outside the United States through joint ventures with European companies, which have led to some of the earlier cash dispensers that Diebold produced and sold in the United States.

Problems occurring in the field did not bring post-mortems or meetings where finger-pointing was prevalent. Meetings at Diebold were called to determine how to correct the problem rather than to blame service, engineering, or manufacturing. Placing blame was subject to severe criticism by Koontz. Managers were to bring management together to solve problems, not to shift blame from one function to another.

Diebold also crafted major technological leap strategies with its products. When Docutel developed a cash dispenser, Diebold supported my recommendation not to come out with a me-too Docutel-type cash dispenser and channeled its efforts into producing a machine that did total banking, which ultimately became the ATM.

When Mosler came out with its TV banking systems, it added some features. But rather than copy these features, Diebold leapfrogged Mosler. For example, because the TV cameras were mounted to the side in Mosler's equipment, the tellers always were looking off in the opposite direction. Diebold responded by developing an "eye-to-eye-contact," half-silvered mirror TV system. When a bank customer pulled up to a Diebold TV bank, the teller, looking from the monitor, was staring the customer right in the eye.

Diebold beat all of its competition by coming out with the first Underwriters' Laboratory–approved data safe, which protected computer magnetic tapes. This fire-resistant safe doubled the time protection being offered by the competition, which had not yet begun designing for electronic data storage.

Innovation paid off for Diebold. In 1965, when I joined Diebold, its sales were $57 million. Today it has expanded to more than $300 million in annual sales and is the market leader in ATMs.

Diebold succeeded because Ray Koontz understood that a company cannot afford to build its culture around a particular product or technology. He knew that change is inevitable and that a company will prosper only if its culture is built around an understanding of the nature of change; he knew how that change affected Diebold's customers and the company itself.

The Cascade of Change

No single phenomenon—not interest rates, markets, inflation, nor even technology—will play as large a role in determining the success or failure of corporations in the future as will their response to change.

We can best describe change as the state in which the future will not flow in orderly fashion from today as we have been accustomed. Change has always been with us, but it was never before as pervasive as it is now. Change is a cascade in business today; one small change leads to a flow of changes through divisions, departments, products, processes, and even with customers.

For example, a tantalum price reduction in Australia lowered the price of tantalum, a metalic element used in Mallory's capacitors, thus reducing Mallory's revenues. Mallory's managers had to make more products to sustain revenues, which required more machinery, more investment, more R & D, and a revision of the firm's plans. A single change in one resource price in the global marketplace caused a cascade of changes throughout several countries, all of which touched virtually every department. This happens every day to companies around the world.

Most managers fear change. Their attitude is that change is at best annoying and at worst destructive. Change wastes money and manpower; it upsets plans, requires expensive contingencies, and makes production lines hard to lay out. In short, change is something that must be coped with by making more changes—unpleasant, but necessary changes.

This anti-change attitude is dangerous. Managers and companies that continue to operate with it will become victims—obsolete, out of business. But companies and managers that learn to incorporate change into their daily operations will become victors instead of victims. These companies will learn to make change part of their business. They will see it as an opportunity rather than a disruption, just as managers at Diebold saw the coming of electronic technology as an opportunity instead of a threat to Diebold's traditional business. Successful managers will recognize change as the raw material of innovation, the milieu in which they must work to prosper. They will learn, in other words, to *innovate change.*

Innovation as a Product

The traditional view of change is that it is a shift in course that occurs between long periods of smooth sailing. Managers must anticipate and prepare their companies for the shift in course. Yet change is more than that: It is a raw material that can be used to craft a useful product called innovation. People have a great deal to do with this process. In fact, people are the bridge between change and innovation. To succeed in the future, managers must learn how to build that bridge over and over again, in every situation. To keep a continuing stream of innovations headed for the marketplace, they must see changes, both inside and outside their companies, not as disruptions but as opportunities they can use.

Ken Iverson saw this when he took over Nuclear Corporation of America, a company specializing in isotopes and nuclear equipment, in 1964. Not only did he see changes in the market for Nucor's products that would be devastating, but he saw changes in other markets and technologies that he could use. He did the opposite of what most companies would do: He innovated by going low-tech. He dropped the nuclear lines and changed the mission of the company to making steel.

Today, Nucor is one of the few remaining specialty steelmaking companies in the United States, and it is a very profitable one. It also continues to innovate new steels and new steelmaking processes, staying ahead of the Japanese and of materials development in other fields.

Nucor's story just proves that the size of a firm or the commodity it specializes in does not matter. Every aspect of every business is changing every day, and each of these changes represents not a disruption but an opportunity for innovation.

The Elements of Change

Change today is characterized by two critical features. First, it usually has more variables than the typical manager can handle at once. Second, it is accelerating so fast that it is uneven, leading to sudden "jerks" of change that can throw even a good manager off balance.

The impact of multivariate change is profound. Traditionally, managers have divided work into disparate segments that flow one after the other, at least so it was believed. For example, the output of an assembly line was considered to be the sum of the time it took for each worker to work on the parts, plus the time it took the parts to go from one machine to another. But today's managers have discovered that the time it takes for a part to move through an assembly line was not the sum of the time it takes to machine it at each station plus the travel time and buffer time in between plus an average for delays. Managers have learned that delays don't average. There are too many variables for a pattern to emerge, and parts were spending much more time waiting for work to be done on them than anyone had guessed.

Change is no longer a single event occurring over a long time span. Rather, change today is multivariate, making it difficult for many managers to comprehend.*

The second critical aspect of change is that it is moving at an accelerating pace, a situation that engineers describe as a third derivative. Physics tells us that when an event moves at an accelerating rate, pockets of change occur unevenly in bursts or, as the mechanical engineers call them, jerks. These jerks are unpredictable, but they come with increasing frequency as the acceleration continues.

Change has always been normal in business, but until recently the rate of acceleration was slow enough so that people could adapt easily. Now, however, these jerks are becoming more frequent, although they still arrive at unpredictable intervals. A major burst occurred at the end of 1979 when oil prices shot up, opening new opportunities for technologies that use less energy. Suddenly, in the metals industry, laser hardening and laser heat-treating processing, once considered expensive, now compete against energy-intensive technologies that had been part of industrial America for over a century.

---

*Studies conducted by Don Genro, AT & T's industrial designer and a leading expert in ergonomics, suggest that people usually can deal with five to six variables at once. The dozens of variables used on a day-to-day basis by business managers are so far beyond these limits that they can easily lead to system breakdown. Information from personal conversations with Don Genero of Henry Dreyfus Associates, New York, NY.

While managers were adapting to the oil shock, deregulation bursts hit the banking, trucking, airline, and travel industries. Then, in the middle of the deregulation bursts, every manager in the nation suddenly was confronted with the AT & T breakup and, along with it, the major changes in all companies' communication costs and operations.

In the past decade the pace of the acceleration of change picked up, jerk by jerk, guaranteeing that every day would be chaos for managers—unless they could understand how to make change an opportunity for innovation.

The Five Critical Domains of Change

Managers can't keep track of all the variables, but they can keep track of the ones that will affect them the most. To do so, they must understand that there are five domains of change that will be critical to managers in the future.

1. *Technology.* As the Diebold story illustrated, every company's business is being changed by technology. Personal computers have turned corporate finance departments into spreadsheet operations, and soon every financial manager will become a computer spreadsheet specialist. Innovations in flexible manufacturing systems and computer numerical control (cnc) machines have so reduced setup time for machine tools that they have deskilled an entire class of workers, machine-tool setup operators.

Of all the domains of change, technology has the most variables and is moving the fastest. Technology's third-derivative pace is all around us, producing jerks of change such as a tenfold increase in microprocessor capacity in a single year.

2. *Demographics.* The aging of the population, the demographic shift to the South and West, the increasing racial and linguistic diversity of our population, the rapidly growing number of working women, all will play an important role in the success or failure of every company in the coming years. Tobacco companies must contend with the increasing awareness of health and fitness. Manufacturers of domestic products must learn how to appeal to women with little time for domestic life. To succeed, companies must follow these demographic shifts and understand how they will change both the market for their products and the labor force makeup to produce these products.

3. *Changing Worker Values.* As every manager knows, shifts in worker values in recent years have revolutionized life on the shop floor. Workers want more involvement in and control over what they do. Quality circles and other techniques to involve workers in decision making in the factory have become almost routine in many major corporations.

What is important to innovating change, however, is not so much what the change in worker values has been but the fact that worker values are changing so rapidly and that, whatever the changes are, they must be factored into a manager's decision making. Adapting to the last change is important; anticipating the next change is crucial.

4. *Changing Laws and Regulations.* According to one estimate, a company like General Motors needs more than 20,000 employees just to keep track of changes in law and regulation that might affect its operations. With regulation, deregulation, and *re*-regulation, tracking these developments can take up a vast amount of managerial time in firms as different as small clothing companies and huge computer corporations.

Yet few areas of change present as much opportunity.

When banking laws were relaxed to allow more branch banking, bank equipment companies such as Diebold, Mosler, and Burroughs exploded with growth because of the sophisticated information systems needed for multi-branch operations.

5. *Global Economics.* The remarkably rapid shift to a global economy is affecting American business managers in dozens of important ways. Managers today must keep up with currency fluctuations, legal requirements in several nations, customs duties and regulations, and even wars and revolutions. Domestic manufacturers must be wary not only of foreign competitors but of their domestic competitors who use foreign markets to build sales high enough to underprice them domestically.

Managers must now consider user preferences and local regulations and skill levels in many countries when they design products. Area managers must keep an eye out for industrial targeting, a strategy gaining in popularity in the "miracle" countries of the Pacific Basin. They must recognize not only the importance of comparative labor costs domestically and abroad but also the overhead—the relative cost of doing business overall. Furthermore, the globalization of the economy is accelerating the pace of technological innovation, making capital equipment obsolete more quickly, and leaping to radical changes in the technological "balance of power."

From Territorial Defense to Domain Thinking

Managers have not traditionally been taught to understand how to deal with the five critical domains of change. What they were taught was the concept of territory. Everyone had their territory—marketing, technical, sales. We all operated within our own territory, taking excursions outside of it when we wanted to tie together somebody else's territory with ours. The dominant activity in corporate life was defending your territory.

In the future, managers are going to have to change their thinking—to turn away from the idea of defending territory and, instead, to learn how to tie all the domains of change together. Diebold succeeded in leading the electronic banking revolution by combining its understanding of several different domains of change. The Diebold management team knew that technological breakthroughs made ATMs possible, that regulatory changes made electronic banking legal, and that changes in demographics and lifestyle had broken down resistance to banking by machine. Diebold no longer looked at the world in terms of the marketing territory, the sales territory, and the technical territory. Diebold knew that, in fact, it was all the same territory and that managers in all areas needed to understand the critical domains of change and work together to succeed.

EMHART'S TRANSFORMATION: INCORPORATING CHANGE

When T. Mitchell Ford became president of Emhart Corporation in 1970, he held the reigns of an old New England company that grew through acquisitions. Ultimately Emhart became the corporation that put together United Shoe Machinery and American Hardware. One of its most important manufacturing lines—glass bottle manufacturing machinery, which had been one of Emhart's original businesses—was under seige in the short term by metal cans and in the long term by plastic bottles. Every market, every technology, and every customer that Emhart relied on was in the midst of a virtually global revolution.

Ford and his senior staff decided they too must start a revolution. So they set out to change Emhart into a firm dedicated to planned, company-wide innovation. Emhart's success is described in the recently published book *The Innovators.* The authors specifically discuss Emhart in a chapter entitled "A Mild Revolution."*

Emhart's R & D managers launched a technical development program to signal that change and new technology would be important throughout the company. They set up a major communications program so that the company's fifty divisions around the world would learn to share the new emphasis on technology and change. Ford created the position of director of productivity, a position responsible for using new technologies to upgrade process and production.

The Board of Directors created a Technology Review Committee with five of Emhart's outside directors to review the company's technology plans and also to ensure that technology was applied throughout the company in offices, assembly lines, and information systems.

Emhart's corporate technology staff instituted a yearly technology innovation award—a plaque and cash prize presented by the CEO to the company's outstanding innovators (with their families present) during a board meeting. Ford and his senior executives initiated a flexible planning process that provides di-

---

*James Botkin, Dan Dimancescu, and Ray Stata, *The Innovators: Rediscovering America's Creative Energy* (New York: Harper & Row, 1984), pp. 56–80.

rection and stability and yet adapts to unpredictable events.

These changes fit well with Emhart's practice of mixing all levels of employees in the company cafeterias—senior executives and clerical workers alike. Whenever possible, CEO Ford and the senior executives ate lunch in the cafeterias and sat and chatted with anyone and everyone.

The changes began to pay off; the once stodgy old company started churning out innovations. In response to the assault on their glass bottle machinery division, Emhart's glass machinery technologists redesigned their machines to make glass bottles cheaper than plastic ones. Emhart decided not to fight the can revolution but to join it by inventing and patenting pull-top aluminum can tabs and licensing the manufacturing rights to their competitors, the can makers.

Furthermore, Emhart proved ready to change with its many rapidly changing markets. When the door hardware market suddenly shifted from an aesthetic orientation to security consciousness, Emhart's hardware division brought out high-security deadbolt locks for homeowners. When hotels and motels began to experience security problems, Emhart marketed sophisticated electronic locks. When electronic components switched from prong configurations to flat chip designs, Emhart's Dynapert Division was ready with an automated electronic surface mounted component placement system. When the automotive industry needed automated stud welding for stronger chassis, Emhart's fastener divisions developed robot-controlled welding tools now in use throughout Europe and the United States.

Other innovations tumbled out: the consumer glue

gun, machinery to mass produce complete circuit boards, and many more, so that in a few years every one of Emhart's fifty divisions across the world was either No. 1 or No. 2 in its market.

As a result of this thrusting innovation, Emhart dedicated more and more funds to developing these new technologies. In 1980, Emhart devoted less than 5 percent of its capital expenditure budget to new technology equipment. By 1984, that figure had increased to more than 50 percent. In dedicating more than $200 million over five years to improvements in manufacturing technology. Emhart was making an unprecedented commitment to the future for a mature corporation.

As a result of this dedication to high technology, Emhart was able to develop state-of-the-art computer-assisted-design/computer-aided-manufacturing (CAD-CAM) systems, flexible manufacturing systems, high-volume automation systems, and robots. The results of this innovation program are impressive: In 1985, Emhart's divisions gained between 5 and 15 percent of their revenues from products that did not exist five years earlier. These revenues were generated by literally hundreds of innovations in products, manufacturing processes, and marketing strategies and techniques.

One of America's supposedly dying mature businesses, Emhart demonstrated that any company can survive and profit in the 1980s and 1990s if it accepts change as the order of the day and thinks of innovation as its product.

2

QUANTUM LEAP INNOVATION

When I joined the Singer Corporation as Vice President of Engineering and Chief Technical Officer of the Sewing Products group in 1971, it had one of the best-known brand names in the world. The word Singer was practically synonymous with sewing machines, and, with more than $1 billion in sales, Singer's sewing products group dominated the global market. Even within the company, employees believed strongly in Singer's identity as a sewing machine company, even though Singer made many other products.

Nevertheless, Singer's sales were in a serious decline in the early 1970s. Home sewing had suddenly become unfashionable. Dresses and skirts, the easiest fashions to sew at home, went out of style, replaced by jeans. And as cheap, ready-to-wear imports became available, home sewing lost its economic advantage as well. Most importantly, although Singer didn't realize it for several years, the women who made up Singer's natural market no longer had time to sew. More women were entering the work force.

At first, Singer responded to this change in the marketplace the way it always had, by adding more "bells and whistles" to its machines. Such additional features had helped Singer fend off the invasion of lower priced Japanese sewing machines a few years earlier. "Bells and whistles" included a vast number of features that could be demonstrated during sales demonstrations in Singer stores. Among these features were decorative stitch patterns such as Greek key, arrowheads, dogs, and trees; attachments; chainstitching; and fabric feeding for difficult materials. Of course, many of these features were beneficial for the customer. This time, however, bells and whistles weren't going to work. By 1974, the drop in sales had become a serious problem, affecting all brands regardless of cost.

With strong evidence that the natural market was declining despite an intensive R & D effort aimed at new features, I decided to find out the real story. Who really was the customer for sewing machines in the United States? And did they really want all those new features or did they want something else?

To my surprise, I discovered that the big push for additional features had come not from the customers themselves but from the sales people at Singer's famous 5000 Sewing Centers, who wanted the machines to have features that would be interesting and easily demonstrable during a sales pitch at the store.

Indeed, when our research teams went out and talked to the customers themselves, a very different picture emerged. In focus groups, the customers said that once they took the machines home, they didn't even use the fancy features that the sales people liked to demonstrate. For most customers, the machines were so complex that they were difficult to use. Thousands of people who preferred the better fit of homemade clothes—Singer's natural customer base—bought ready-made clothes instead because it was just too hard to figure out how to run the sewing machine.

A change in the market had gone virtually undetected, and then that mistake had been compounded by a new product features program aimed in absolutely the wrong direction. The net result was a disastrous plunge in sales. Singer knew that it had to change.

The solution was a quantum leap in innovation: The Athena 2000, the first electronically controlled sewing machine ever put on the market, was introduced on June 1, 1975. The Athena had been developed in complete secrecy over a four-year period at the sewing products group's headquarters in Elizabeth, New Jersey. In bringing the Athena to market, Singer ignored all of the home-sewing industry's traditional sales and marketing patterns.

Instead of making sewing more complicated by adding mechanical features, electronics would make the Athena less complicated. The 500 mechanical parts in the old Singer machines were replaced by a single microprocessor. Any one of twenty-five different stitches could be selected simply by pressing one button. Instant reversal and automatic buttonholing could be selected by pressing another button. The instruction manual was cut from sixty to twenty-four pages. The Athena was the first user-friendly sewing machine.

Instead of listening to its sales organization alone, Singer developed the electronic Athena machine. The marketing program—directed at the user, not the sales staff—stressed simplicity, speed, and the ability to sew modern fashions with new fabrics.

And instead of attacking costs and price against competition, Singer developed a highly proprietary, top-of-the-line product. Believing that customers would pay more for a versatile, easy-to-use machine, the company gave the Athena a $1,000 pricetag, making it one of the most expensive sewing machines ever sold.

This quantum leap strategy may have turned the conventional wisdom upside down, but it also saved Singer's top-of-the-line market segment and forever changed the design of sewing machines. Singer sold more than 1 million Athenas—generating over $1.5 billion in revenue for the company—before any other manufacturer could put an electronic sewing machine on the market.

Prior to the Athena there was a shifting trend toward low-cost and simpler mechanical machines. With the introduction of the Athena, Singer revitalized top-of-the-line sewing, providing customers with sewing machine simplicity far beyond what could be achieved by mechanical machines. With the Athena, Singer had not simply innovated a product; it had innovated a whole new way of thinking about home sewing, both inside the company and among its customers. The quantum leap strategy had worked.

Making a Quantum Leap

Most managers today think of innovation as synonymous with product extension—the process of refining a product and adding features to make it more competitive, as Singer did before the Athena. In most cases, the added features do not affect the operation of the product but merely its position in the market. It is still the same product, competing in the same market for the same customers.

But, as the Athena story proves, to compete in today's marketplace, managers must change their thinking entirely so that innovation becomes more than just bells and whistles. To make a quantum leap, managers must think not in terms of mere product extension but in terms of managing innovation for success.

Quantum leap strategies examine the product, the competition, and the market carefully, not in today's terms, but in tomorrow's. They project where the market is going to be in 5 or 10 years and what products or services will place the company at the top of that market. Then, profitable ways to take advantage of this changing marketplace are designed.

Sometimes existing products are used as a starting point, but frequently the managers must completely rethink every aspect of *the company,* not just the product, to gain the leading position in a market. This means innovating throughout the firm to achieve a market goal, just as Singer had to do in reorienting its sales staff for the Athena. Indeed, in some cases the product itself may not change at all. Instead, production, service, sales, or even financing is innovated in a way that puts the product far in front.

Bypassing the System

The usual strategy for product introduction involves a first roll-out. This is a stage in which the bells and whistles are added and the product line is diversified to hit different market segments and more refinements are made in response to competition. Eventually, the product's sales level off as the market reaches equilibrium and is saturated. Finally, a new model is brought out or the product is killed by a new technical development or a shift in the market or consumers' tastes. While the product is undergoing refinement, the manufacturing process itself undergoes refinement, lowering the price and extending the product's life. The combination of these two processes can take years.

To quickly build additional features into your product, you could crouch in the bushes, watch for the competitor's weakness, and then pounce with a machine that has an extra feature or two. This is not a bad strategy if it could be achieved within a relatively short period of time. But from my experience additional bells and whistles bring their own debugging problems, the new introduction takes longer than planned, and the competing product gets a major market share in the meantime. Meanwhile, the competition begins to develop its own bells and whistles, negating your research.

But in order to compete in a world where change is the rule rather than the exception, managers have to learn to bypass ordinary design and development delays. That's what quantum leap strategies are designed to do. Under a quantum leap strategy, as soon the first product is rolled out, the company begins planning for the new technology or market shift that will supercede it. Sufficient numbers are sold to pay back research costs and tooling, but stress is continually placed on developing the next technology, not only on refining the existing one. Refinements may be added to meet competition, but the thrust of research and development is aimed at new breakthroughs, complete with innovations in processes, manufacturing, sales, and service.

Again, this requires a change in the company's mindset. But need not be a change that threatens the company or its managers; in fact, it can be liberating. For example, if your company uses processes that result in scrap, you are probably working on methods to reduce the scrap rate to save money. Try a quantum leap strategy of researching *forming* processes to completely eliminate scrap and produce better shaped parts in the first place rather than working on improving *cutting* processes that might reduce scrap later. Within Emhart, manufacturing operations managers continually explore cold forming processes to produce parts that are very close to final specifications rather than cutting processes that produce the part and its associated scrap.

If your company is working to obtain incremental gains in quality by reducing the number of rejected parts, try a quantum leap program that examines every operation and corrects any defects at each work station to ensure that no part leaves a workstation with a defect. At Mallory in Indianapolis, the emphasis on incremental gains through quality control has been supplanted with an overall program that ensures making the parts right in the first place. The program has made Mallory the top appliance control manufacturer in the area of reliability.

If your company is attempting to refine its existing technology to solve a problem for your market, try applying a technology from another market. Diebold leaped over its competitors in methods used to detect bank vault break-ins by burglars who used a high-temperature tool known as a burning bar to burn through vault walls. While its competitors were trying unsuccessfully to develop metals that could withstand burning bar attacks, Diebold installed highly sensitive smoke detectors inside the vaults, which signalled the slightest penetration by tools that use heat.

If your company is struggling with a quantum leap technology that does not yet do the job, don't give up. If it holds the promise of dominating the market because of fundamental values it holds, someone will get the bugs out and make a fortune with it; it might as well be you.

If your company is doing well with an existing technology, but your reading of the critical domains of change are that you will need to move quickly with shifts in markets or technology in the future, leap over your own technology before some other company does.

Creating a Quantum Leap Strategy

Over the past thirty years, I have learned a number of useful techniques for generating quantum leaps and bringing the results successfully to market. All of these techniques have one thing in common: They require that managers go far beneath the surface of today's situation, digging deep into the future to find real ways to innovate not just Band-Aid solutions. I will return to these techniques over and over again in the course of this book, but let me list them for you right now:

1. *Ask the Innovation Question.* Never sit back and think that you have finished innovating. Always ask whether there is some lower cost, quicker, and more efficient way of doing any of the processes you are involved with.

2. *Design and Market for the Result, Not the Product.* Don't find yourself married to a particular technology or a particular way of thinking. When you ask the innovation question, always ask it in such a way that you can leap over your own production processes or attitudes if that is what you find you should do.

3. *Get to the Root of the Problem.* When confronted with a problem, look past quick fixes to what I call the 124 percent solutions—solutions that deal with problems at their most basic level.

4. *Make Innovation Second Nature.* Don't just tack a few innovative ideas onto a staid underlying corporate environment. Overhaul your company in whatever ways are necessary to create an environment that will produce innovation on a managed and systematic basis until it becomes second nature.

Asking the Innovation Question: Chet Carlson's Copying Machine

"There must be a quicker, better way of making these copies!" The voice was that of Chester F. Carlson, a patent attorney with the P. R. Mallory Company, a New York electronics firm in the 1930s. Carlson was constantly exasperated by the tedious process of making duplicates of drawings and patent applications. The only method existing at that time was a slow, costly photostat process.

His colleagues always agreed, but they also always noted that nobody had ever found one, so they continued using the photostat machines. One day he shot back, "Maybe nobody has ever tried" and decided to do something about it.

And that's how Chet Carlson asked one of the most important innovation questions of the century.

Carlson had come to Mallory's patent department after earning a B.S. degree in physics from the California Institute of Technology. He soon realized he needed a law degree to advance in the patent department, so he enrolled in a night course at a New York law school. He used public library textbooks, copying a great deal of the information by hand. The resulting writer's cramp spurred the thought that there must be a quicker, easier way of copying. Soon, in both his work in the patent office and his nighttime studying, the innovation question became his obsession.

His training as a physicist and his exposure to battery research at Mallory led him to explore a method of printing called electrophotography. Mallory was then conducting research and development in batteries, and Carlson had read reports on the company's related processes. The various electrical characteristics of the powders in the dry cell gave Carlson ideas on how a xerographic process could create an electrostatic image from toner powders. He began research in the kitchen of his small New York apartment, filling it with glass slides, jars of chemicals, resins, and lamps—equipment that cost almost all his savings. He focused on the idea of using photoconductivity for taking pictures of documents electrically to eliminate the slow, wet development process of conventional photography.

On October 22, 1938, Chet Carlson produced the first electrostatic copy. On a glass slide he inked "10-22-38 Astoria"—the date and the location of his experiment. He rubbed a cotton cloth over a sulphur-coated metal plate to give it a static charge, placed the slide over it, and exposed it to a glare from a blazing floodlight. He then dusted the metal plate with a vegetable-based powder. When he blew the surplus powder away, the inscription was visible on the metal plate. When he pressed a wax-coated paper hard against the plate and peeled it away, it was imprinted with a recognizable transfer of the lettering on the metal plate. The xerographic process was born!

In 1940, Carlson obtained the first of many patents. During the next four years he continued to try to interest other companies in his new invention, but to no avail; more than twenty companies turned him down. P. R. Mallory himself decided that his company could not do anything with the process because it was heavily committed to improving the dry-cell battery. But Mallory permitted Carlson to pursue the patent on his own. Ultimately, Battelle Institute in Columbus, Ohio, entered the picture, bringing together Haloid, a dying photographic company, and Chet Carlson. No doubt you've guessed the outcome. The invention led to the company—Xerox.

Because Carlson had asked the innovation question—"How can this be done better, faster, cheaper?"—xerography was born. Because P. R. Mallory and a long list of other prestigious corporations did *not* ask the question, they missed the chance to capture an innovation worth billions of dollars.

Designing for the Result, Not the Product: Electronic Locks

For thousands of years, almost everyone in the world has used mechanical keys and locks for security. But as hotels and institutions have encountered problems with lost or stolen keys, electronic locks have become increasingly popular.

At hotels the situation is difficult. Security control is easily lost because guests, wittingly or unwittingly, walk off with keys, and many levels of employees, from managers to housekeepers, must have master keys.

For years, lock companies had solved the problems with mechanical locks by using more sophisticated mechanisms. The solution was to provide different levels of key security. But Russwin & Corbin, divisions of Emhart, were also able to solve the problem by marketing electronic locks.

Electronic locks allow each door to be programmed to accept a certain code, contained on a "key" that resembles a credit card. Thus, a hotel can change the code every time a new customer checks into a room or whenever one of the card keys is lost. By the same token, housekeepers and other personnel can be given only limited access; a master key that works on one floor may not work on another, for example, and the keys can be recoded regularly to improve security.

The combinations are endless, as are the opportunities for using an electronic lock in many ways other than simply providing access. For example, the time and identification of the entrant can be provided for computer rooms in industry or research institutes. And, at the same time, Russwin & Corbin is able to retrofit hotels and motels with electronic keyholes that are aesthetically pleasing and provide the company with a large new market.

Russwin & Corbin's step forward was not really the introduction of electronic locks. It was the ability to see innovation in terms of the result rather than the product. The competitors believed that people expected keys in hotels and would not accept cards. But Russwin & Corbin saw that people use plastic cards to activate hundreds of devices in common usage and would not resist electronic locks. Again, to succeed, a company found that it had to rethink not only the product but the customers and the marketing approach as well.

Getting to the Root of the Problem: The 124-Percent Solution

Here is a problem any manager without the quantum leap way of thinking is likely to encounter. If any department tries to produce 100 parts, only 80 will be made well enough to pass quality assurance inspection. The remaining twenty are returned to the system for rework or reinspection or are sent back to the vendor.

I have seen this situation in literally hundreds of departments and processes during close to three decades of management. It is accepted as the norm in almost every U.S. company or factory I have visited as well as those in several other countries: 100 parts go into a process, 80 good ones come out, and 20 are rejected or reworked. Thus, to get 100 good parts, a manager will often manufacture 124.

This 124 percent problem holds many American firms back from innovating and competing. A commitment to quality would seem to be a prerequisite for success in any business, yet it really is a quantum leap for many firms. And oftentimes, even when they *think* they have devised a quality strategy, they haven't really made a quantum leap.

Once managers begin to operate on the assumption that they must order 124 parts to get 100 good ones, the entire organization soon develops a 124-percent mentality. Everyone in the company shifts to working so that it can effectively make 124 of something, knowing that only 100 will come out.

Interestingly, this results in everyone working hard. Expeditors are set up, new positions are created so that parts can be moved, and when it comes time to target cost reductions, the organization works on the 124-percent level. Cost reductions are achieved on how best to deal with the 124 parts, not how to produce 100 parts.

For example, if quality inspection is a major problem because of the 24 defective parts, personnel in the 124-percent system will cleverly devise ways to carry out inspections more effectively. Inspection tools will be recommended, statistical screening processes will be set in motion, and a quality control department devised with extra inspectors to seek out the 24 defects.

I can recall a case in Singer's plant in the United Kingdom where powdered metal bushings were placed in machines to reduce the need to oil Singer machines in the field. These powdered metal bushings were impregnanted with oil under pressure as they were produced. When shafts rotated in these bushings, a self-lubricating effect took place, leading to machines that did not need oiling, a great market advantage.

During the production of one lot of machines, however, the bushing had not been properly impregnated with oil. For some reason, the process did not apply the oil under pressure. Rather than reject the whole lot of bushings, thus delaying the production of machines, it was decided to hand-oil these bushings with oil cans. On the production line, this proved to be a very effective way to move the machines past quality control, into shipping cartons, and out to the field.

The proud manager who resolved the problem was credited with keeping that production line going by recommending the oiling of these bushings by hand.

The only problem was that when the oil-less bushings had been introduced, Singer's service organization had been advised that it no longer had to oil the particular places where the bushings rested. Thus, as the machines hit the field and ultimately the oil that was placed by hand evaporated from the system, certain parts began to freeze up. As they froze up, the service organization attempted to free up shafts by putting in new subsystems, never realizing that that particular lot held bearings that had been defective in the first place.

The 124-percent mentality had struck again. Everyone on that production line was proud of the fact that they had solved a very tough problem. But the solution cost the company thousands of dollars and incalculable inconvenience because of service problems later on. The managers of this production line had learned how to operate in the 124-percent environment. What they had not done was figure out how to create a 100-percent environment in the first place, that is, to correct the process that made the defective bushings in the first place.

Here lies the secret of what Japan has accomplished over the years. Japanese managers have simply recognized that the most effective way to operate is to order 100 parts manufactured and, if all 100 parts are not usable, investigate the reason why. They have found that the true 124-

percent solution is not to figure out ingenious ways to get 100 good parts out of 124 made but to foster a true 100-percent mentality.

Making Innovation Second Nature

Innovation and management processes in the United States are moving toward more creative environments, even among the more prosaic manufacturing firms. "Freedom to fail," "managing by wandering around," "intrapraneuring," "skunk works," "bootleg" development money, risk taking, cross-breeding and multidisciplinary teams, allowing for self-selection of new product developers—these and other techniques for generating and nurturing creativity have been well chronicled and are starting to become standard elements of American business culture.

Freedom to fail, risk taking, and all the rest are indeed necessary for innovation in design and development, but they aren't enough. At some point, a quantum leap has to make money; it has to sell, and it has to sell better than anything else in its market. But how can these techniques be used to successfully bring to market a steady stream of breakthroughs that use quantum leaps to bypass the normal delays in product and process refinement?

The answer, although it is not easy, is simple: Innovation must be systematized. Products and processes must be continually revolutionized to stay ahead of the competition. The process is ongoing; it is encouraged throughout the company. It is this method of managing innovation that I will discuss in the next chapter.

3

MANAGING INNOVATION

On a cold winter day in December of 1940, the executive offices on the fifty-third floor of the RCA building in New York echoed with angry shouting. General David Sarnoff, the chairman of RCA—the man who had brought radio sets to a mass market and organized NBC, the first radio network—had just learned that archrival CBS had successfully completed its first experimental transmission of color television earlier that day.

At stake was one of the richest prizes in business history: the creation and domination of a market that would eventually encompass the entire nation and influence communication technologies around the world. The ultimate revenues from the sale of color broadcast and reception equipment would reach well into the billions of dollars; but, more than that, the company that solved the color puzzle would literally change an entire industry, the television industry, and, with it, the entertainment customs and purchases of people around the world.

The General was mad, and he knew what he wanted: RCA was going to innovate its way to leadership in color television broadcasting or else. RCA scientists and engineers at the company's research headquarters in Camden, New Jersey, quickly whipped a crude color camera into a working unit and rushed it to the NBC studios in New York, along with a few television receivers quickly modified to receive color. Soon, Edward Land of the Polaroid Company and other inventors were also working on the problem, trying to beat the giants. The race was on, and it was clear that whoever understood the nature of this historic change best would be the winner.

World War II halted the color television experiments, but, by the time they started up again in 1945, the battle lines had been drawn: In the labs at RCA and CBS, two separate—and incompatible—color broadcasting systems were evolving.

Although I didn't join RCA until 1952, when I helped develop color broadcast studio equipment that made RCA the world leader in the field, I had a ringside seat for the early strategies and struggles as a student at MIT in the 1940s. As a lab assistant to renowned color physicist Arthur Hardy, I was in a position to hear blow-by-blow descriptions from many of RCA's visiting engineers and scientists.

CBS was working on a mechanical rotating disc system. A spinning disc in front of the camera of the TV studio would be synchronized with a similar disc in front of the TV set's picture tube in the home to provide the color. The quality of the color images it produced were bright and clear, but the television pictures could be received only on rotating disk television receivers. The CBS color television pictures could not be received at all on existing black-and-white television sets. By contrast, the pictures being produced by RCA's all-electronic system in the 1940s were of poor quality, pale and scattered, but the television picture could be received on existing black-and-white receivers, in black and white of course.

Yet there was a difference at RCA. Sarnoff and his engineers were working not just on a technological breakthrough but on a vision of the future, and it was that vision that made the difference in the color television contest. Sarnoff, because of this vision of the future, understood two important facts that William Paley, his counterpart at CBS, did not seem to see.

First, Sarnoff saw that to be successful, any color broadcast system had to be compatible with black-and-white sets that were already in use or face tremendous resistance among consumers. CBS was so intent on bringing out the first color television system that it was willing to cut off thousands of existing television viewers to do it. Meanwhile, excited RCA engineers were showing Sarnoff color and black-and-white television sets, sitting side by side, receiving the same broadcast, an impossibility with the CBS system.

Second, while Paley and the CBS engineers saw only the high-quality color picture on the sets they were developing, the RCA engineering team saw that the rotating-disc system had reached the limits of its technological development, while RCA's electronic system had boundless room for improvements. CBS was simply not prepared for the changes that were likely to occur in the years ahead; its engineers believed that a good electronic system, if it was possible at all, was far in the future. RCA not only saw those changes but relished and fostered them. In other words, CBS's engineers and managers had allowed change to box them in; RCA's had learned to turn change into an opportunity worth billions of dollars. RCA's management team saw the quantum leap innovation.

But there was more to RCA's success in the color television race than merely understanding change; RCA managers also undertood how to act on that knowledge by planning and managing innovation in an entirely new way. For the color television offensive, RCA marshalled the considerable forces of one of the most technologically advanced and innovative corporations in history and launched the R & D version of the Normandy invasion. Thousands of employees were organized and directed in a serious, programmed, and planned effort to overtake CBS, involving not just scientists and engineers but technicians, marketing specialists, lawyers, and managers in dozens of laboratories and offices throughout the country.

Although CBS management knew the stakes the two corporations were playing for, in the final analysis, RCA proved to be more innovative. Despite the power and technological resources, CBS committed a series of strategic and tactical errors.

In 1946, for example, CBS tried to force the issue too early by seeking FCC approval of its embryonic technology. The FCC denied the approval simply because the technology was not ready. If CBS had waited and recognized the future of an all-electronic technology, the story might have ended differently.

By contrast, Sarnoff knew exactly how to use the legal domain and other nontechnological areas to his advantage. In 1950, CBS dealt RCA a serious blow by winning FCC approval for the rotating-disc system. But RCA appealed the FCC decision to the Supreme Court and rallied both its engineers and the public through speeches and advertising around the concept of color broadcast that any television set could receive. (RCA had also sold several million black-and-white sets during the battle with CBS that could not receive CBS's color signals.) But most importantly, RCA's management team understood and communicated the urgency with which RCA had reacted to the technological change forced upon it by CBS, or be forever shut out of the market.

The following year, RCA gave design and production data on its compatible all-electronic color system to CBS and 231 other manufacturers, shifting legal and public interest away from CBS's system and encouraging all television set manufacturers to get in on what was obviously going to be a technological revolution, thereby winning over not only the public but television set manufacturers, including RCA's competitors. Finally, on December 17, 1953, the FCC adopted the color TV specifications that put the RCA system on the air. RCA color sets incorporating its new tri-color television tube, as well as black-and-white receivers picking up the color broadcasts, all performed marvelously. Sarnoff was aglow with triumph, and NBC immediately began to broadcast commercial programs in color.

With the development of color television broadcasting, RCA proved that innovation is not restricted to products; it must encompass everything connected with the company. Nothing should be left out of innovation—not customers, not the industry as a whole, not the government regulatory agencies, and especially not the competition. Along with the idea that innovation is all-encompassing goes the idea that it is continuous. Innovation cannot be a one-shot deal; to help a company survive in a world of change innovation must be planned and managed to provide a steady stream of breakthroughs that find a profitable position in the marketplace. The color television victory did more than merely sell color televisions. It also helped establish within RCA a culture of innovation.

Few companies that failed to learn this lesson in the postwar years are still in the same business today. While CBS became a factor in television programming, it was never a factor in the manufacturing and marketing of home television receivers. RCA became the dominant domestic supplier of television sets to the consumer. After thirty years of applying the lesson I learned in 1952 when

RCA's all-electronic color pictures reached the public, I can say with confidence that no company that fails to learn the RCA lesson will still be in business at the turn of this century. Such companies will be the victims of change, just as CBS was in the color television battle.

Managing Innovation in the Real World

As the story of color television technology shows, RCA is one of the most innovative companies in electronics. Yet RCA has had a number of failures over the years—most notably its disastrous entrance into the computer market in the 1960s. Other companies experienced defeats even though in certain market areas they were highly successful. Addressograph-Multigraph was heralded in the 1960s as an example of an excellently managed company, yet two decades later it was restructured under Chapter 11 of the Bankruptcy Act.

In the real world of innovation there are few unmitigated success stories, just a lot of ups and downs. One part of a company does well, another part does not. Some products work, some do not. Even in the most innovative companies, managers confront situations that block their creativity and confound their executives. The ultimate fate of IBM's PC Jr. and Osborne's breakthrough portable computer show that, in the realm of innovation and excellence, even market leaders and trailblazing newcomers with reputations for excellence can fall victim to unevenness.

What we learn from the real world is what distinguishes companies that excel and innovate despite their environments: Success comes from within, not only from the business climate, the state of the economy, or any other outside factor.

Companies that are "real world innovative" remain so over long periods of time. They have strategies that enable them to cope with both internal and external ups and downs, balance off the setbacks with new opportunities, and demonstrate a consistency of performance.

Real-world innovation recognizes that a firm must produce a constant stream of breakthroughs so that doing so becomes a habit. This is the way it can offset the inevitable failures. Real-world innovators plan for phenomenal growth but accept failures and mistakes as inevitable lessons. They combine the freedom to create with the systems to produce, market, improve, and phase out products and introduce new ones as the critical domains of change shift.

Real-world innovation is a track record of breakthroughs that stretches over time. The 1985 shakeout of many legendary Silicon Valley companies is evidence that innovation in the long run is more than good ideas in the labs; it is the ability to innovate change throughout the company consistently and profitably over time, moving the innovation from the lab into the marketplace.

Innovative companies may gain prominence with a bold new product that captures a tremendous market, but they remain successful through a flexibility that allows them to change with the shifts in markets and economies. Their managers know that no matter how good they are now, they will face potentially disastrous changes in the future—and they are ready to handle those changes.

Real-world innovation built Xerox, which grew out of a troubled Haloid Corporation to become one of the world's giants. It saved United Shoe Machinery, which survived a disastrous antitrust suit. The classic case of real-world innovation is, of course, Chrysler, America's success story of a company that virtually rebuilt itself and its entire line of products to survive in a very competitive marketplace.

"Stretch" Planning in the Real World

Companies that have successfully endured recessions, inflation, stagflation, currency revaluations, global revolution, and legislative and political upheavals have done so by fostering a climate of innovation and success, particularly among their managers.

These managers know that building and selling an innovative product is a great start, but to succeed in the marketplace in today's world, they must continue to build and sell their innovative products over time with an increasing return on investment. What separates these managers from ordinary ones is, first, desire and, second, an understanding of what I call "stretch" planning.

Innovative managers are not content merely to produce a product that is successful today. They want to play an important role in expanding the company and improving its performance. If they can do so, they will see their company as one that fosters an environment oriented toward innovation and opportunity, and they must foster that environment throughout the entire company.

But innovative managers do not make such a contribution just once in their careers or by happenstance. They contribute to the *continual* improvement in the company's growth *by design.* They *plan* to grow at phenomenal rates, and they manage their company not for the next quarter but for constant increases in the rate of growth for five or ten years. They stretch!

Many managers and executives in basic businesses assume slow, steady growth keyed to business conditions and the state of the economy. But a new wave of firms beginning in the "Silicon Valleys" of the nation and moving outward have established new standards for growth. Companies like Compaq ($105 million in its first year and over $300 million in revenues in its second year) have redefined what is possible. The entire concept of corporate growth has been affected by the third derivative pace of change and is itself changing.

Innovative managers will set *stretch goals* to expand the growth of the corporation for the five-year projection and then begin thinking about how the company can remain victorious in twenty years when its current products have given way to products that don't exist now. They set targets of the percentage of sales they will achieve from new products in five years and at how much research and development they will have to do to achieve this growth curve yet be flexible enough to change as conditions change around them. And they don't merely look for 5 percent growth; they look for 20 percent and more. This requires that the managers continually plan for a future in which 20 percent of their company's yearly sales come from products that do not exist today.

Once a company begins to approach its growth plan from that perspective, its managers are forced to keep it in mind all the time—to look at what they have to do this week, this month, this year to reach those phenomenal stretch goals years down the line. If change is the raw material for innovation, then successful managers need to master how to convert this raw material into finished goods. They must convert the opportunities of change into the innovations of the future.

To successfully manage innovation, future managers must know their customers, their competition, and themselves.

PART II

Know Your Business

4

KNOW YOUR CUSTOMERS (THEY MAY NOT BE WHO YOU THINK THEY ARE)

When General Sarnoff and his management team laid out RCA's strategy for an electronic color television broadcasting system, they knew that the pathbreaking planned-innovation approach to research and development would not be enough to guarantee success. They also realized that technical innovation was useless unless the results could be sold in the marketplace. Furthermore, they understood that RCA's customers were not just the people who would eventually purchase color television sets. They knew they had to sell three sets of customers, one after the other, and change their business strategy for each one.

The first customer was the Federal Communications Commission, which had to be sold on the superiority of the RCA system before Sarnoff could even begin to think about introducing it into the marketplace. The marketing strategy for this customer was to convince the FCC that RCA's system was better than CBS's because even though its picture quality was inferior at the time, RCA's color broadcasts could be picked up on black-and-white sets already in use.

RCA's next customer was the broadcast industry. They had to persuade television executives that the investments made in the RCA system would pay off in the short run, because advertisers would want to reach viewers with the same commercial whether they owned black-and-white or color sets, and in the long run, because the RCA system would be safe from mechanical obsolescence.

Then the final customer, the viewer, had to be sold the television sets—a fairly routine exercise in introducing and merchandising a new consumer product.

Sarnoff and his lieutenants carefully targeted their sales efforts to each customer in turn, switching from comparison selling with the FCC, to targeted client sales with the broadcasters, to mass-market merchandising with the viewers. Had they not done so, any of the three customers could have failed to buy. The color system would have had no chance for success in the marketplace, and RCA's innovative approach to research and development would have been useless. RCA made electronic color television a success by understanding who its customers were, even when they were not simply the people who purchase television sets.

Who Is Your Customer?

The bottom line for innovation in business is meeting the customer's needs. You can always create new products, but unless you know who your customers really are—unless you know to whom to sell those new products—you will fail as an innovator because you won't be able to make new products profitable. This may seem like an elementary lesson in business, but, incredibly, many businesses don't really know who their customers are.

Singer did not know who its customers were when its engineers added more features to its mechanical sewing machines. The sales organization wanted them, but the sales organization wasn't the customer. The real customers, home sewers, wanted a simpler machine. Singer was innovating for the wrong people.

The Workslate Computer developed by Convergent Technologies was a marvelous piece of technology and a real breakthrough in computer engineering, but the prospective customers, finance professionals, did not perceive a need for it and did not buy it, even though it was designed specifically for them. Convergent Technologies incorrectly assumed that because the customers were using other equipment to do what Workslate could do faster and more elegantly, the customers needed the new product. But the customers used the same equipment for many other operations, too, and saw no need to buy a second piece of equipment to do what their current computers were already doing. Workslate innovated for the right customer but the wrong need.

Diebold, on the other hand, built a very profitable banking service and equipment empire by knowing what banks and bank customers' needs were, sometimes before the banks themselves knew. It came to the banks with full-service banking machines instead of a cash dispenser that duplicated what was already on the market. It created drive-up windows for neighborhood banks, recognizing the future branch banking needs. Diebold's secret is simply that it always keeps in mind who its customers are, and it carefully monitors the five critical domains of change, not only to see how change will affect its own operations or its own technology but how it will affect the future of its customers.

A COMPANY THAT DIDN'T KNOW ITS CUSTOMERS: THE ADDRESSOGRAPH-MULTIGRAPH STORY

Addressograph-Multigraph was a company that based much of its success on the sales of its multilith copying machines to corporate duplicating departments. Its customers were the duplicator managers of those departments—or so it appeared.

When Xerox's instant copiers first sought to invade Multigraph's market, the duplicating department managers and the Multigraph sales force reported that Xerox's machines were not a threat. With the help of the duplicator managers, salesmen smuggled imperfect Xerox copies out of customers' offices to prove how poor their quality was. Reports to management were filled with tales from the "customers" about the problems of Xerox machines. When the market research department surveyed the duplicator managers, they were told ovewhelmingly that Xerox machines were too expensive for the duplicating departments.

But the real customers for duplicating machines were not the duplicator department managers. The real customers were the secretaries who had to put up with the inconvenience of the multilith duplicating system. The secretaries wanted a Xerox machine in the corner because it was faster and more convenient than typing a master copy of a memo and sending it to the duplicating department for delivery later in the day (or later in the week).

The duplicator managers who *appeared* to be Multigraph's customers did not want to believe that Xerox was a better system for the secretaries, because instant copying in every office would reduce the need for their central duplicating departments and would be more costly. And the Multigraph sales force did not want to believe that Xerox was a threat either. They had been indoctrinated with the sales-oriented philosophy of their feisty septuagenarian CEO, J. Bass Ward, who always said, "We have got to keep those multilith duplicators moving."

In the end, however, the corporate executives who decided which machines to purchase did not listen to their duplicator managers or to the Multigraph sales staff. They listened, logically enough, to their own secretaries. No matter how many innovations Multigraph's managers added to their multilith duplicating machines, they could not stop the Xerox invasion. But they might have been able to if they had known who their customers really were. Two decades later, the reorganized AM International recognized its market as medium- and long-run duplicating. Today AM knows its customers.

Innovate for the User and Renovate the Sales Force

As the preceding examples suggest, the customer for an innovative product is not always obvious. Ironically enough, it is often the sales department that makes it difficult to see who the customers are and what they need, and the sales force may need more convincing than the actual customers themselves. But it is possible to chart paths of opportunity through the organizational and structural mazes in order to get to the customer. Three principles can help keep innovations on the right track.

First, *design for the product's ultimate user rather than for the sales staff or the research staff.* The best innovations come about because a company knows who will use the innovation and how. If the sales staff offers "cultural" resistance, then the sales culture must be changed because there are always competitors inside and outside the company who can meet the ultimate user's needs and sell the right innovation to the right customer.

Second, *market your products to an entire industry. Don't sell them just to the people who buy them.* Diebold and other successful suppliers study their customers' businesses as carefully as the businesses themselves in order to anticipate every need. Again, the sales people can be a formidable obstacle here because, ironically enough, they do not understand who the customer is. Harvard's marketing genius, Ted Levitt, sums it up by advising that "selling focuses on needs of the seller."*

Levitt suggests that face-to-face selling concerns itself with the tricks and techniques of getting people to exchange their cash for your product. The result is a narrow product orientation rather than a broader customer orientation, which leads to neglecting nonproduct elements of the sale such as service and refinements. Thus, a strong salesperson may be able to close a particular deal, but ultimately the market will shift to a company that understands the importance of developing a position within an industry based on more than product. Without this position, even the most agressive sales force will fail.

---

*Theodore Levitt, "Marketing Myopia," *Harvard Business Review* 53, no. 5 (September-October 1975): 31.

Third, *don't count on intermediaries for marketing data.* Companies whose products are transferred through intermediate steps need to do their own market research on the ultimate user. This is important especially for companies that are OEMs (original equipment manufacturers). Very close working relationships with the intermediary's marketing people should be set up so that they become part of the research and evaluation process. Better yet, co-marketing or independent marketing research and data sharing should be provided for in the initial supply contract.

This holds true even for companies that readily share data with their suppliers. Diebold found that banks were very sophisticated in understanding their customers' needs and the needs of their markets, and that banks were also willing to share that information with their suppliers who would work very closely with them. Diebold set up its management structure to take advantage of this fact by giving the product managers for each of its major product lines authority to establish close working relationships not only with the banks' executives and corporate staffs but also with branch managers and the banks' customers.

OTHER EQUIPMENT MANUFACTURERS: CATCH-22 FOR INNOVATORS

One mechanism that sometimes stifles innovation in American business today is the other equipment manufacturer (OEM) structure. OEMs are arrangements under which one company supplies another with components or products that are then sold to a final user, frequently with the buyer's name on them.

Craftsman power tools and certain Kenmore vacuum cleaners, for example, are manufactured by the Singer Company. Singer supplies the products to Sears, which sells them to the public. Most automotive parts are sold under OEM relationships, as are many home applicances. Large retailers like Penney's use hundreds of OEM suppliers for everything from televisions to electric blankets.

Component suppliers in OEM relationships often do not have direct contact with the end users of the products they build. They must depend on the companies they supply for the feedback on products necessary to innovate.

Sometimes the buying companies provide their OEM suppliers with all the market data they have, but it is too general to be useful to product innovators in the OEM firm. In other cases, market information is not available to the OEMs, either because no system exists to collect and transmit it or because the buying company feels that consumer data is proprietary and therefore secret. In these situations, the supplier firms—the OEMs—are discouraged or even prevented from going out and getting it themselves. As a result, the designers and builders of thousands of products are denied the information needed to innovate.

Furthermore, when the market information is available, and when the OEM uses it to develop new products, it is obligated to take the idea only to its own customer rather than to the ultimate buyers. This traps the OEM in a dilemma. If the OEM does test its new product with the ultimate buyer, the OEM's customer will complain that its turf has been invaded, jeopardizing an important business relationship. But if the OEM takes the idea only to the retailer and it is rejected, the innovators don't know if the customer doesn't like the idea because of territorial reasons or because the market data argues against it.

Paradoxically, one of the suppliers' competitors or a firm from an entirely different field can bring a new idea to the OEM's customer using its own market data and sell the customer its innovation. Competing suppliers are not obligated to bring ideas only to the one retail customer; thus the competitor, even if it does not have experience with the product, does have the information and the ability to innovate. Thus, while the OEM is wrestling with this dilemma, its competitor can steal the customer.

In the past few years, several OEMs and their customers have learned to deal with this problem. Firestone Tire and Rubber, for instance, set up its own market research system for the tires it sold to major auto manufacturers, and the auto companies came to realize the value of such a system. Similarly, Singer established its own market research system for the products it manufactured for Sears and, as a result, was able to bring Sears such innovations as the Shop Vac. Sears realized that the market-specific research developed by its OEMs was very useful not only in the line Singer provided but also in marketing their other floor care products.

A variation on the OEM dilemma can occur as a result of the information flow between divisions of a single company. When RCA's television receiver division was developing home television sets, its planners concluded that no set with a screen larger than ten inches would ever sell. This conclusion was based partly on a rule of thumb that related viewing distance to the diagonal screen size, but there was a second reason as well: The engineers in the division believed that a television tube more than ten inches in diameter would also make the set so much larger in depth that it would be too large for most living rooms.

RCA's tube division did not believe this line of reasoning; market researchers there conducted their own survey and found that Americans wanted as large a screen as they could buy. As a result, RCA Executives Elmer Engstrom and George Brown ordered the tube division to develop a set with a screen greater than ten inches but that was not appreciably deeper than sets with smaller screens. Experts in electron optics solved the problem, producing a major innovation in the television industry—an innovation that might never have come about if RCA had listened to its own experts instead of its customers.

Knowing the Customer-Product Loops

The best way to understand who your customers are is to visualize your company as a wheel with a hub in the center and five spokes radiating out from the hub. The hub is your customer. The key centers in your company, where innovation occurs, are located along the outer ring of the wheel. The spokes are "loops" along which information flows between the customer and the key centers in your company. (See Figure 4–1.)

Visualize the spokes as two-way communication loops with your customers: They link different parts of your company with different aspects of your market. The *marketing loop* contains information on the product's mar-

Figure 4–1. Customer-Product Loops.

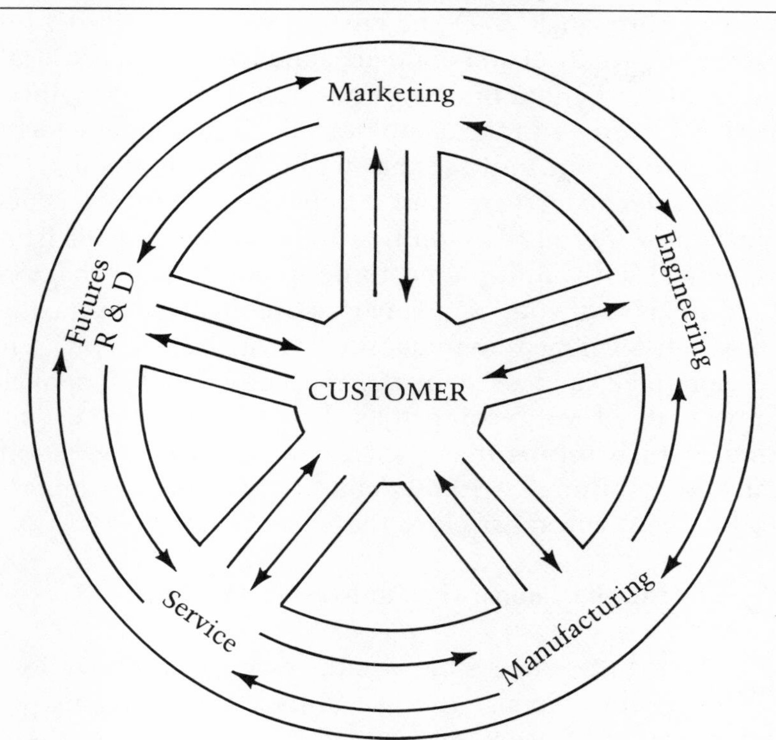

ket, its needs and competitors' plans and products. The *engineering loop* contains information on the design and technology needed to develop the product. The *manufacturing loop* contains the customer services and manufacturing processes needed to maintain quality control and to best meet the customer's needs and delivery requirements. The *service loop* carries to and from the customer information about the product's performance and reliability, customer support needed, and the quality and availability of parts and service. The *futures* or *R & D loop* provides information pathways for a combination of functions that examine through R & D the customer's future needs, including information and dialogue about the product's legal and regulatory environment, future refinements or adjustments for changes in the customer's market, and new technologies to be put in place that will affect the product.

In order to prepare your company so that it accepts change as a way of life, you first must visualize your firm as a wheel with multiple customer-product loops that include all areas of your company that are involved in getting a product developed, manufactured, distributed, sold, serviced, refined, and used in the marketplace. The spokes connect all of your company's departments to the customer, which means that no department is able to go off and "do its thing" without considering the customer's needs. Then, you must close the loops.

Closing the Loops: the Innovation Potential

Closing the wheel's loops is the essence of innovation. By closing these loops, you can unlock a tremendous potential for innovation within your companies. If these loops are functioning, your customer's needs will be used to guide and stimulate innovation throughout the firm. Each loop brings information into the company that stimulates innovation in other departments. I've seen this happen, and I know how people react: They get excited about what they have learned. The excitement is contagious, and innovation begins to race around the firm.

Customers are also recognizing the potential of direct links with the spokes of the wheel. Increasingly, they are insisting upon dealing directly with the manufacturing operations and demanding that their suppliers pay the same attention to the supplier-vendor network. As a result, new loops are formed between the customer and parts of the company that have never had contact with the customer. This generates new products and services keyed to a customer's precise needs.

A sign of the growing demand for direct communication between end-user and manufacturer is the proliferation of toll-free (800) service numbers. Popularized by computer and software firms, 800 numbers are now used by companies such as General Electric to provide customers with direct contact to the factory.

Once you have learned who your real customers are, you can follow four steps to close the communication loops and realize the innovation potential they hold.

First, *use the loops to link innovators with customers.* Diebold used its sales structure as an effective conduit from the customer to its product developers. This gave Diebold direct contact with thousands of branch bank managers around the country. Diebold could do this because of the marketing culture that CEO Ray Koontz built into the sales force.

Singer's best links with its customers were the sewing course instructors who taught new buyers how to sew at home. Not only did the instructors relay complaints and questions directly back to engineers and designers, but, because they used the machines in virtually every conceivable situation, they became an excellent source of innovative design ideas themselves.

Addressograph's service organization played the same role in the development of credit card systems. Not only did they report problems customers had with existing Addressograph systems, they also alerted the firm about the first use of plastic card systems by stores. And they were able to help the engineers select the type of equipment needed for a large-scale introduction of plastic credit cards to all of Addressograph's customers.

By systematically collecting these reports, Addressograph's engineers and technicians were able to stay ahead of the competition in innovation and keep their customers very satisfied with back-up service. Addressograph found the service staff so valuable a source of information that service personnel were assigned to the teams that developed the charge systems in use today worldwide.

R & D managers, product development managers, manufacturing engineering managers, and marketing managers responsible for implementing new products and services become successful innovators when they can tap in on two-way communication links to the final users. When products are sold, these links can be created through the company's marketing and sales organization; later, they can come about through manufacturing and service links. But it is the ability of the innovative manager to establish two-way links that ensure that innovations meet the right needs for the right customer.

And don't forget your internal customers. Companies operate through a vast network of informal channels interconnecting many diverse company functions. Innovative managers learn how to use these internal informal channels to their best advantage. They learn to tap into or help establish channels for two-way information exchanges between departments and divisions inside the company so that their needs can be met by innovators throughout the company. Establish a two-way communication network between the product and system development people and the rest of the company.

Second, *wander around the market as well as the plant.* The term *managing by wandering around* has become a popular phrase within innovative organizations. But managers need to expand the idea by wandering around their customers as well.

Wandering around the customers means more than asking them questions, taking surveys, and holding a few focus groups. If customers have never heard of a product or a process, they will not know whether it has future potential. So wandering-around managers must infer customers' needs from complaints they hear as they wander. IBM has systematized this wandering around the market by getting its executives out into the field on sales calls on a regular basis.

Third, *don't run the product relay race.* In most firms, the product design process amounts to nothing more than a relay race: Information from the customer is collected by the sales and marketing people and then handed off to the development lab, which works on it and hands it off to the engineering staff, and on and on until the first widget rolls off the line and is shipped—often to a distributor who is not sure who will buy it.

In the relay race, information usually flows only one way, from the customer to a single department. Then it gets interpreted, modified, memoed, and handed to the next department, complete with modifications, delays, biases, and mistakes, all of which compound as the development goes through successive departments and levels of the company. When it finally reaches the manufacturing department in the form of drawings and process sheets, it may only remotely resemble what the customer needs, even when careful attention is paid to market research data and product specifications.

The problem is that sales, marketing, engineering, and manufacturing are viewed as separate skills rather than as part of a unified system. Data is gathered and presented like a race baton to be passed along by contestants. Unfortunately, the baton—the specifications for the product the customer wanted—changes at each handoff.

Your firm can avoid the relay race by closing communication loops between your customers and all of the firm's departments. This requires managing the company as if each department were part of an integrated customer-serving system made up of a series of loops each linked with your customers.

Finally, *study the customers you can't see—they can hurt you.* Companies that sell to one or two types of customers often do not diagnose the needs of other potential customers. Even successful firms often fail to understand that if they do not continue to innovate for these other potential customers, new firms will move in with innovations and new services that may ultimately penetrate and dominate the company's traditional markets. The American steel can industry was invaded by the producers of aluminum and plastic in this manner, and now aluminum is being threatened by plastic and paper products that have been popular for years in Europe.

Companies that rely on domestic markets alone will miss the foreign market potential for their products. As competitors go after the expanding foreign market and learn to satisfy those market and user needs, they will threaten entrenched domestic suppliers. These unseen customers pose threats and offer opportunities. Managers who extend their sweep of the five critical change domains into new markets can avoid the former and take advantage of the latter.

Once you know who your real customers are and what their needs are throughout the innovation process, you can start studying who your competition is. Like everything else, it has changed since last you looked.

5

KNOW YOUR COMPETITION (IT MAY NOT BE WHAT YOU THINK IT IS)

The Friden Company, which in the 1960s was a leading manufacturer of mechanical calculators and mechanical office products found it difficult to successfully enter the electronic era. In the 1970s the original Friden, which ultimately became part of the Singer Company, ceased to exist. Like many traditional mechanical product companies, Friden failed to recognize the early warning signals of electronics looming on the competitive horizon.

As a vice president of Diebold, I had the opportunity to observe how Friden management underestimated electronics as a threat to their business. Diebold's check imprinting division used Friden Flexiwriters as data entry devices for personalizing checks. The Flexiwriter was a leading mechanical product used throughout business and industry as a data input device for many systems applications.

With the increasing volume of bank checks, the mechanical Flexiwriter was rapidly becoming obsolete. Thus I initiated discussions with Friden management to alert them of the need for a computerized electronic system to replace their Flexiwriters. What we encountered at Friden was the belief that mechanical improvements to the Flexiwriter would sustain that product in its market-leading position.

Throughout my discussions at Friden, its management continued to emphasize the virtues of the Flexiwriter over other competitive mechanical products. They refused to accept electronics as a threat to the traditional Flexiwriter, a product widely used throughout business and industry. Today, very few Flexiwriters continue to be used; they were replaced by systems other than those manufactured by Friden or its mechanical competitors.

Friden's managers thought they knew who their direct competition was: other mechanical office equipment and mechanical calculators. Likewise, you may think you generally know what, or who, your direct competition is: the other companies that make the same products or offer the same services you do. But, as the Friden Company found out, you may not recognize your competition even when it is placed right there in front of you.

Competition no longer means simply Ford and Chevrolet. Today, competition means anything that diverts your customers' attention, and ultimately their money, from your product. It can mean new technology that will make your product obsolete, new lifestyles that make your product irrelevant, or new regulations that make your product illegal or impossible to market. Like virtually everything else in the business world today, competition is getting harder to keep track of. It is accelerating at an accelerating pace. And it is more varied than ever before, coming from more directions and more places than managers even five years ago had to think of. Understanding your competition is a necessary step to out-innovate it.

Although competition today comes in all forms, we can narrow it down to two basic types: direct and indirect. Direct competition means a product or service that provides the customer with essentially the same thing that you are providing. But even direct competition is not simple to detect these days, as the Friden experience points out. Their direct competition came from microchips, even though Friden thought the direct competition was from other calculator manufacturers.

Indirect competition is the competition you are not looking for, the competition that comes from an unexpected arena. It is the fastest growing variable in the competitive environment today.

Many of Singer's problems in the late 1970s arose because the company did not understand the nature of its indirect competition. The home sewing market was in a decline, and Singer executives did not look deeply enough to find their true competition. In one year declining sales were attributed to bad weather. The next year the oil crisis was blamed. The following year cheap foreign machines were made the scapegoat. Sales declined every year.

It took a careful market study by Jane Lockshin, Singer's corporate strategic planning analyst, to point out that the real competitors in the U.S. market for sewing machines were the employers who were hiring women in record numbers for jobs previously filled by men. By the late 1970s, women held more than half the jobs in the United States.

Thus, Singer's competition was not Sears, Brother, Bernina, or Huskvarna, nor manufacturers who might seek to supercede existing sewing machine technology by a breakthrough, as microchip manufacturers had done to Friden. Rather, Singer's competition was everything that competed for women's reduced free time: jogging, golf, sailing, stereo, political activism, cable TV, gourmet cooking. The source of pride in homesewn clothes was replaced by other sources of pride, such as physical fitness.

Innovative managers understand that the nature of competition has changed. They know there are new weapons at their disposal to fight the competition they identify, and they know they must spot competition before it can do any damage.

Direct Competition

In the past, managers were always taught to meet their direct competition head-on. If price were a factor in the marketplace, then you beat the competition by lowering the price. If features were the deciding factor, then you developed more and better features. The rule was: Sell against the competition.

But this is no longer the case. Not only do you have to beat the competition in technology today, as the Friden case illustrates, but you have to learn to use, and defend yourself against, today's weapons that are gaining prominence in the global market and be on the lookout for new ones that arise continually. The most important new weapons in direct competition are delivery schedules, quality, and service.

Delivery Schedules

In the 1980s, the U.S. automotive industry adopted "just-in-time," or JIT, inventory practices, requiring suppliers to deliver materials and components onto the factory floor just prior to the time when they are needed on the assembly line. This was a major change for the industry, not only because it cut costs and manufacturing time, but because it required the virtually complete restructuring of production facilities.

Companies supplying the automotive industry either must meet these ultra-tight delivery schedules or become victims of change. Those that meet the schedule will have mastered the use a powerful new weapon in the competitive game: delivery when needed.

The reasons the industry is willing to pay premium prices to suppliers for precise delivery times are not hard to fathom. It eliminates inventory, reduces assets, and cuts manufacturing time and therefore labor and overhead costs.

The JIT concept is now spreading throughout industry, leaving in its wake those suppliers who do not revolutionize their manufacturing, warehousing, scheduling, and inspection processes so they are able to deliver parts to the beginning of their customers' assembly lines hours or even minutes before they are required.

Using these weapons is not easy. It requires a painfully thorough restructuring of assembly lines, inspection systems, packing, shipping, and handling procedures. A major part of promising and delivering to meet JIT requirements is making sure that there will be no problems or rejects. Backup systems have to be set up so that they can come into play immediately in the event of a problem that could affect the delivery. Once a firm misses a delivery date or delivers materials that do not meet specifications, it is permanently out of the running, at least for that customer. The risks are higher and the stakes are higher, so the care taken must be much greater. It is certainly changing the competitive arena for many companies.

Quality

One of Xerox's favorite folk tales, which is told to suppliers during plant tours, involves a plant manager who usually ordered components from domestic firms and was repeatedly disappointed by a high defect rate. He decided to order ten thousand of a certain component from a Japanese firm and stressed that he wanted "no more than five defective pieces." When the order arrived, he noticed that it came in two boxes: a large one containing the ten thousand components he had asked for, and a smaller one containing five more.

When he asked the supplier's sales representative why this had happened, he was told that the small box contained the five defective parts he had asked for. With much apology, the Japanese sales rep said he hoped the defects were the ones the manager had wanted, explaining that since the company doesn't ordinarily make defective parts, they weren't sure what kind of defects they should introduce. The rep added that making defective parts was very expensive and that the company would have to charge extra for them in the future.

In case the suppliers did not get the point of the story, Xerox spokespeople told them that from then on price would be a secondary consideration in awarding contracts; quality would be the deciding factor.

Similarly, Dr. Eldred Rickmers, a nationally known quality-assurance consultant hired by Xerox, told that company's suppliers that Xerox was ending its practice of maintaining contracts with several suppliers. Instead, the company would begin buying strictly from one or two companies that could meet very stringent quality requirements, after meeting those requirements, they would then have to meet the cost targets.

Quality is becoming increasingly important in business, largely because of the invasion of quality foreign goods. It began with the Japanese quality-based competition that overwhelmed the U.S. auto, machine tools, consumer electronics, and light truck industries. The introduction of JIT virtually eliminates the time that could be used to deal with defective parts on the assembly line because the tight delivery schedules will not bend for breakdowns or even for the inspection of parts as they are delivered. The entry of quality-conscious Japanese manufacturers into markets ranging from construction to consumer goods, often with lower overhead and labor costs, has raised the stakes in the quality competition.

Quality, however, is not a result of more or better inspection. It is a reconceptualization of how products are made. Like the Japanese plant manager who had to make the defective parts separately, managers must change their operations so that defects are not just caught and corrected but never made in the first place. This requires tighter tolerances on machine tools, rigorous training and motivation programs for workers, tougher standards for suppliers, maintenance schedules designed to keep machines running perfectly, and constant testing at each step of manufacturing. Quality is the ability to understand the old saying: "We never seem to have the time to do it right, but we always seem to have the time to do it over again."

Service

Competing on the basis of service is not new, of course. But what is new is the understanding that service in many industries may be as important as the product itself.

As I explained in Chapter 1, Diebold first earned its reputation for banking by building vaults that survived the Chicago fire. But it continued to reinforce that reputation through an extraordinary service strategy. Diebold set out from the beginning to create a service organization that met the needs of an industry that handled money—and, after all, time is money. Today, any branch bank manager anywhere knows that a Diebold service engineer will respond *immediately* to a phone call for any kind of problem with Diebold equipment. This service policy has convinced bank after bank that any cost difference between Diebold and its competitors must take into consideration the reliability that Diebold's service guarantees. Diebold recognized that to be competitive in its market niche, service was the paramount competition weapon.

Caterpillar Tractor Company competes on the same principle, guaranteeing parts delivery anywhere in the world within hours, from a worldwide network of parts plants and warehouses. Federal Express guarantees to deliver a piece of mail by 10:30 a.m. the next day or immediately find out where it is. AT & T will fix phone equipment quickly for any customer who leases—a policy that has persuaded many customers to keep leasing from AT & T rather than buy lower cost equipment from other manufacturers. Software companies prominently advertise their 800 service numbers to give immediate help to a customer who may have only misread a manual. One firm, LifeTree Associates, maker of the Volkswriter Deluxe word-processing system, has its vice president for corporate accounts call commercial accounts—even small ones—to take care of any problems immediately. IBM's service on computer hardware is legendary, as is Xerox's service on its copying machines.

While these companies compete on price most of them are priced toward the top of their market. Many of them also compete on product features, but their strength is often their service. They have recognized that in a world of rapidly accelerating change, reliability and service reduce the variables and slow the perception of change in their customer's minds. The service they provide is part of the real product they sell, and it often provides that competitive edge.

UNDERSTAND YOUR DIRECT COMPETITION

You must know your direct competition in order to out-innovate it. You must predict your competition to out-innovate it before it reaches your market. But you cannot do this through seat-of-the-pants observation; you need to set up a structured analysis of potential competition in order to stay on top of everything.

1. *Know your competitors' products—inside and out.* Set up a formal structured competitive analysis acitivity within your organization. Charge someone with the responsibility to "tear down" competitive machines, not only analyze the cost but also assess the performance of your competitors' products. Bring your R & D, engineering, manufacturing, and marketing personnel into the competitive analysis department regularly to observe how your products "stack up" against competition.

At the Singer Company I had the competitive analysis activity "life test" Singer machines against all of its competitors: ranking Singer machines with competitors to determine "out of carton quality" "infant mortality" (machine failure rate within the first hour), reliability, and customer performance.

At Diebold, my competitive analysis activity was managed by an ex-CIA employee, who attacked competitor night depositories, safety deposit boxes, and security files (in Diebold's laboratories, of course), comparing Diebold's security products' vulnerability with that of its competitors' products.

Above all, your competitive analysis must be set up as an independent function, not as part of design; there must be an unbiased culture pervading that activity. I also have discovered that such activity is an excellent training ground for younger design engineers, who can start their design career in the analysis of competitive product features and competitive designs prior to joining a design function.

2. *Know the competition's products in actual practice.* This sounds like the oldest advice in the world, but a remarkable number of people in business world don't heed it. You need to understand how the competition's product works, how it performs at the customer's location and, most importantly, how it is perceived by the customer. Bench tests, videotapes of products in operation, disassembly, and reassembly are all used to greater or lesser degrees by most managers and are critical in an era of rapidly innovating competition.

In cases where it is impossible to study the competition's equipment because of cost or size (i.e., large machine tools or presses), computer simulations or on-site videotapes can be used. Having your engineering staff interview your competition's customers or using your marketing and service staff to interview can provide the kind of information needed to spark innovation in products that you cannot bring into your lab and test. And it is an excellent way to learn how customers perceive competitors' products.

Your competitive analysis function also should follow consumer reports, reviews, and advertising about your competition. Again, these are all activities done to some extent by most firms, but they need to be systematized and placed in a single job function to make sure that the insights gleaned from them reach every corner of your company.

3. *Don't underestimate your competition.* Don't count on your competitors to fail when you hear reports that they are in trouble. And don't count on the failure of an innovation that threatens your business. Such attitudes are likely to lure you into a false sense of security.

Instead, assume the worst: Your competitors are going to correct the problem and come after you. Assemble a what-if task force of people from various parts of the company to speculate on what would happen to your firm (and to their jobs) if it *did* work. Ask them what would happen if it worked even better than your product does now. This can help to engender the innovation environment you need in your plant and also to second-guess the competition. It will also generate new ideas that leap past the innovations your competition may have added.

4. *Learn the competing culture.* Naturally, your competition will downplay your innovations, and its employees will resist cultural changes, just as yours will. But you can learn to take advantage of these weaknesses if you make it your business to know what they are. For example, if your competitors are wedded to a certain technology, will they lag behind if you introduce a new technology? Don't always attack your competitors where they are the weakest. Attack their strengths. Diebold attacked Mosler in electronic security systems, an area in which Mosler was once a

major factor. In telephone line security, Diebold made its competitors' products obsolete by using new technology.

Indirect Competition

The story of Xerox's takeover of the duplicating market from Addressograph-Multigraph, described in Chapter 4, reveals some of the hidden threats of competition; in hindsight it is easy to see that the Xerox machine was in direct competition with AM's machines, but AM was focusing on the threat from other brands of duplicating machines.

The Xerox defeat of AM reveals, however, another facet of indirect competition. The arrival of the Xerox machines hurt many industries besides duplicating, some of which may never have realized that they were competing with copying machines. One was the carbon paper industry, which offered a relatively simple product line that was the major market area for a company like Emaloid Inc.

During the 1960s, Emaloid devoted considerable R & D effort to developing innovations in the use of carbon paper such as carbon paper multiforms; pressure sensitive, encapsulated carbon paper; no smudge carbonless paper; and single-use carbon paper. All of these innovations were aimed at its direct competition, other carbon paper companies. But Xerox virtually killed the market for single sheet carbon paper and drastically reduced Emaloid's revenues in all other uses. Yet, Emaloid could not see the competition from copying machines; sales just continued to drop as secretaries got more copying machines and used fewer carbons. Worse yet, Emaloid could do nothing about it but cut prices and add features that competed head on with other carbon companies.

The banking equipment supply industry also suffered a major loss to an indirect competitor in the late 1960s. Diebold and Remington were battling hard for the lion's share of the new market for mechanized files—huge moving file cabinets that allowed a clerk to store and retrieve thousands of records in large vertical cabinets much faster than they could be retrieved using the old hand method and horizontal drawers.

The two firms outfeatured and undercut one another back and forth to gain what they saw as a market with enormous potential. They soon discovered, however, that they were no longer competing against one another. Indirect competition had entered the field in the form of the information miniaturization techniques developed for computers. Mechanical file manufacturers suddenly realized that the market they were innovating for was being quietly stolen by microfilm, microfiche, and computers. Ultimately, laser disc storage will make obsolete even those products that outdated the mechanized file.

As these examples show, indirect competition is frequently the result of technological innovations. But this is not always the case. Singer was forced to innovate because of indirect competition caused by value and life-style changes. Tobacco and distilled spirits companies are facing indirect competition from the health and fitness value change. The heating oil industry in the Northeast competes head on with coal and natural gas, but in the long run its most devastating competition is the population shift to the Southwest, a region that uses less heating energy overall. The fire alarm industry faces indirect competition from police departments, which have decided to charge citizens for false alarms or even to deny connections to alarms because of faulty signals.

The examples are almost endless. To avoid becoming yet another example, however, managers must continually scan not only their business environment but all five critical areas of change to get early warnings of indirect competition. Two useful methods managers can use to help them do so are what I call "early-warning systems" and monitoring.

Early-Warning Systems for Managers

Indirect competition is unpredictable, but it is possible to create early-warning systems that alert you to possible trouble over the horizon so that you can focus on it and decide if it really will threaten you.

The process is not difficult. These systems probably already exist in your company in bits and pieces. The key is to systematize and coordinate them. Then you can scan the five areas of critical change for signals of indirect competition and routinely subject those signals to "signal processing" in the company to ensure that the information is useful and is used.

A method that we have created uses a formal scanning task force that meets at regular intervals to examine data from a variety of sources that can tip us off to competition emerging outside of our usual domain.

Each member of the task force is charged with the responsibility of bringing to the group's attention any concerns or early signals of change that are occurring in a particular area. Each has developed systems to scan the media, other companies, scientific journals, and other sources for trends that could offer new markets, or compete with the company's existing products. Decisions can be made on the spot to expand investigation or even to begin research into counterstrategies.

This also eliminates the problem many firms face when they turn the task of scanning for new trends over to the R & D departments: a narrow focus on technical areas that interest only the research staff. With the corporate scanning task force, indirect threats from nontechnological areas as well as technological ones are more likely to receive notice.

I have also initiated forecasting operations to draw up "what-if" scenarios about events and trends in society that could affect our products. Forecasting, however, is not always valuable in dealing with information from fields outside the primary markets of the company, usually because the information is weak or because the people involved are not equipped to go outside of their specialties. Nonetheless, sometimes forecasting contracts to think tanks and universities can provide crucial insights that lead to innovations far ahead of the competition.

Monitoring

The monitoring method that has been most effective for me is espoused by MIT Professor James Utterback, head of MIT's industrial liaison program, along with James Brown.*

In principal, Utterback reduces the definition of monitoring to four straightforward activities:

- searching the environment for signals that may be forerunners of significant technological (or other) change;
- identifying the possible consequences (assuming that these signals are not false and that the trends that they suggest persist);

*James M. Utterback and James W. Brown, "Profiles of the Future," *Business Horizons* 15 (October 1972): pp. 5–15.

- following and verifying the true speed and direction of technology and the effect of employing it through well-defined parameters and policies; and
- presenting the data from the foregoing steps in a timely and appropriate manner for management's use in decisions about the organization's reaction.

Where I have used this technique, the results have been most revealing. For example, at the Singer Company these efforts helped identify the early signs of declining home sewing. Changing demographics shifted women away from sewing to other leisure time activities as more women entered the workforce. At Diebold we foresaw the microchip creating a technological revolution leading to new electronic products for banking where companies like Diebold applied silicon chips to products that were ultimately leading to electronic funds transfer.

The critical areas of change are challenging managers as never before. Not only must future managers identify these impacts of change, but future managers must, most importantly, foresee the time period when these changes will impact their businesses.

At the time that I joined Diebold, it was predicted that electronics would change the banking industry. For these predictions to materialize, however, required the further development of the microchip, which then led to the development of ATMs. Further, branch banking legislation required clarification before remote banking systems became accepted by banks nationwide.

Similarly, at the Singer Company, punched paper tape driven sewing machines in the 1960s met with no commercial success. It took the development of the microprocessor in the 1970s, over a decade later, to permit Singer to develop their highly successful Athena 2001 consumer electronic machine.

I discovered early in my management career that it is not sufficient to merely identify the changes that will affect your business; the real key to management success is to predict the timing of these events. Introducing a product or process too soon into the marketplace may be disasterous since the marketplace may not be ready for the innovation. For example, RCA was way ahead of its time in introducing automation to the newspaper industry in the 1950s. Both newspaper management and labor resisted technological change. The structure of the printing industry was predominantly letterpress printing with a culture of engraving and hot metal typesetting. Within this culture, RCA was ahead of its time with electronic automation.

A decade after RCA's newspaper automation failed, the structure of the industry changed to increased offset printing. Management and labor became more receptive toward new technology and other companies, not RCA, succeeded in applying the technology that RCA had introduced prematurely.

The thrust of this scanning, in Jim Utterback's words, is to look for "straws in the wind" indicating change and ask how best to evaluate them and at what point to take action. Utterback proposes that we must look for indicators that reveal combinations of economic and technical events, as well as political and social forces that may either create demand and support for a new product or process or create resistance to a new product or process. He says we must also look primarily outside the firm to industries and technologies other than the firm's current competitors.

Now that we have looked at our customers and competition, we must examine ourselves. We may not be who we think we are.

6

KNOW YOURSELF (YOU MAY NOT ALWAYS BE WHO YOU THINK YOU ARE)

All competition is not external. There is, in a sense, another form of competition, a form that is harder to control. That is the competition that grows within your own company. Whenever you try to identify indirect competition, or use technological advances or other methods that will allow you to confront direct competition, the first battle you will have to win is within your own organization. Because they always feel more comfortable with the status quo, your employees will probably respond to innovation in one of the following ways.

Our way is better. This was the response at Singer when electronics first began to invade the sewing machine and office equipment markets. The people who "grew up" in Singer were convinced that mechanical principles applied to sewing or typing would not be replaced by electronics. Indeed, a new technology has an uphill battle within your own company. Watch the change, however, when competition and your customers accept and use new technology.

It will never work (here). This is a variation on the first response. In the late nineteenth century, messenger companies in London refused to become telephone companies because there were plenty of messenger boys to deliver messages. Other innovations that were rejected by certain firms because they could "never work" include the propeller, the printing press, the railroad, and, of course, Chet Carlson's photocopying technology, which was rejected by the many companies who stayed with their traditional copying methods.

Why change? The Farrell Company, a division of Emhart, had established dominance over the market for an industrial polymer batch mixer, the Banbury mixer, so that it was the standard worldwide. When the company introduced a product that used an entirely new method of processing polymers, consuming less energy and doing a better job, the sales force discovered that longtime customers were reluctant to change. They felt the new process would have to be proven before they would change to this new technology. Thus, even your customers can be reluctant to change.

Clearly, each of these three approaches simply will not work in a world where the only constant is change. No way is ever better forever, and no product is ever so dominant that it will continue to be profitable forever. Yet when they are confronted with change, most managers at most companies will respond with one of those three protests.

Why? Because their corporate culture prevents them from viewing the situation any other way. A corporate culture is, in the words of Stanley M. Davis, "the pattern of shared beliefs and values that gives the members of an institution meaning, and provides them with the rules for behavior in their organization."* In other words, corporate culture is the ambient background of tradition, symbols, roots, attitudes, perceptions, values, codes, philosophies,

*Stanley M. Davis, *Managing Corporate Culture* (Cambridge, Mass.: Ballinger Publishing Company, 1984) p. 1.

myths, power, and principles—proclaimed or latent—that pervade an organization day after day. Over time, this culture affects the way a company's employees view everything they see or do in their work.

Depending upon the culture's basis and its treatment of change, it can be a powerful instrument working for or against innovation in a corporation today. Diebold's Ray Koontz understood this when he deliberately built a corporate culture based on meeting the needs of the banking industry.

By contrast, the culture at Addressograph-Multigraph was so wedded to the multilith duplicating machine that managers at the company could not see the Xerox threat. This culture was shaped largely by the Addressograph-Multigraph management team, which constantly reminded everyone that Xerox could not overtake offset duplicating. Perhaps this was good for stimulating sales, but it was detrimental when such an attitude pervaded strategic planning meetings.

Other companies also let their duplicating technology subcultures block innovation. The 3M company had a Thermofax copying culture based on a heat-sensitive paper. Kodak was selling copiers based on a photographic copying system Verifax, which sometimes left the copies moist and damp. The Charles Bruning Company sold diazo equipment to copy engineering drawings, a process which frequently smelled of ammonia. In spite of their shortcomings, each company defended its particular product and refused to consider innovating a xerographic product.

Corporate Subcultures

Like the culture of a city or nation, the culture of a corporation is not a single culture but an amalgamation of many cross-cultures and subcultures. There are executive cultures, management cultures, labor cultures, technology cultures, marketing cultures, and even customer cultures. All play a key role in a company's innovation. New cultures are constantly emerging within business and industry today, partly as a result of the rapid pace of change. A case in point is the totally new personal computer culture that has appeared within virtually every company in the nation. Unlike their predecessors, who understood and used only mainframe computers, executives and managers today are becoming comfortable using personal computers at work or even at home.

The overall culture of the company is the background against which dozens of subcultures and sub-subcultures play, vying for attention and resources. Within any firm you can find a labor subculture, a legal subculture, an engineering subcluture, a sales subculture, a marketing subculture, and many others. The members of each frequently do not realize they are part of a subculture. If you ask them to describe their company's culture, they will describe the world view of their subculture within it. If you offer observations from other subcultures inside the company, they will play down the validity of those views. Even executives, who presumably have a global view of the firm and should be aware of all the subcultures within it, frequently do not realize the extent to which they work within a narrow subculture.

Internal subcultures can clash. The Singer Company found that when it went into defense contracting, the sewing culture within the firm was often overwhelming, causing serious morale problems within Singer's newly acquired military electronics companies.

Singer also made high-precision gyroscopes, Craftsman Power tools, and other nonsewing products. But its sewing culture decreed that Singer's real business was sewing products—the business that had built the company—and the new divisions found great difficulty in being assimilated into the company.

WIN ONE FOR THE NIPPER: RCA'S CORPORATE CULTURE OF CONSTANT INNOVATION

When I went to work for RCA in the early 1950s, the company's research and engineering offices in Camden, New Jersey, weren't much compared to the luxury work environments in the Silicon Valley today. They were red brick buildings with cornices and pillars that overlooked railroad tracks and the vast industrial activity along the Delaware River. During our eight o'clock rush to work every morning, a slow freight ran down the tracks in the middle of the street past the building where color television studio equipment was being developed. During the summer, the entire complex was surrounded by an endless line of trucks hauling tomatoes to the Campbell's soup plant nearby.

In the shadow of the tallest RCA building visible from the Ben Franklin Bridge connecting Camden and Philadelphia was RCA's logo: "Nipper," the famous cocked-eared, spotted fox terrier. The little guy glowed like a beacon over the labs that developed many of the pioneering innovations in radio and television.

In fact, Nipper was everywhere at RCA—on calendars, in paintings, in advertisements, on record labels. There was a Nipper Grill adjacent to the cafeteria where one could order a Nipper Special (a club sandwich). The company store even sold Nipper salt and pepper shakers.

Nipper is not only the world's best-known dog, but also the world's most widely recognized and most enduring trademark, registered in the U.S. Patent office in 1900 by the Victor Talking Machine Company, an RCA predecessor. Francis Barraud, an English artist, captured his dog's fascination with voices emanating from an early Edison photograph. The resulting picture, "His master's voice," became an RCA legend and we were proud of it.

We may not have realized it at the time, but we became part of a corporate culture every bit as pervasive as that created by company flags and songs in Japan. We experienced the aura of technical excellence that permeated the entire company and invigorated the management. Technical breakthroughs were the basic source of earnings growth, and scientists and engineers at RCA had the internal power. The company slogan—"The most trusted name in electronics"—was our badge of honor, and Nipper was our mascot and inspiration.

But Nipper was only part of RCA's corporate culture. The rest of it came from General David Sarnoff, RCA's chairman. Sarnoff was the catalyst behind the company's remarkable commitment to innovation in electronics.

Sarnoff used to stimulate breakthroughs by asking his people for birthday presents—specific technical innovations to be delivered by his birthday. The we-can-do-anything scientific and technological culture at RCA was so strong that Sarnoff's birthday requests would set in motion thousands of scientists and technicians racing to create the next electronic marvel, sometimes with unexpected and humorous results.

One year, at a lunch celebrating his anniversary in broadcasting, he took the podium and requested three birthday presents for his fifty-fifth birthday. The presents he asked for threw RCA's R & D labs into turmoil for years: a light amplifier, a refrigerator with no moving parts, and a way to record and playback television.

Scientists at the luncheon snickered at the first request because it would violate the laws of physics, but they still went to work on it, and their research eventually led to the miniaturization of television cameras, a development that totally changed television news. No one found a use for the multi-ton refrigerator that had no moving parts, but the third request ushered in the videotape revolution. And even though that present was initially delivered by AMPEX, a competitor, it was RCA that applied videotape to color television playback across the country.

The culture that produced these marvels was so strong that RCA's management was willing to be innovative about funding research for the General's presents. One time, for example, Sarnoff issued a bold challenge to the RCA Astrophysics Military Group, which was working with Lockheed and Kodak on a secret project to develop a satellite-based, high-resolution photographic system that would send film back to earth from space using parachutes.

The General stood up before us all and said, "Someone is going to develop a television camera in space transmitting these pictures to earth, and the first camera had better be an RCA camera."

The team threw together an unsolicited proposal to

the military for a space-based television system. Knowing that the pricetag for such a venture would never receive military approval, we gave it an artificially low price, figuring we could bootleg money from other parts of the company to pay the difference. The technology we developed because of the General's challenge has enabled the world to see astronauts, the surface of Mars, and the rings of Saturn in its living rooms.

Electronics was the lifeblood of RCA—a way of life. When we visited customers to explore their problems, in our minds we already knew that electronics would be the solution. We used electronics to inspect beverages inside bottles and to measure the uniformity of a mass of hot molten steel. RCA color television was used for teaching complex surgery at Walter Reed Hospital by setting up color television cameras that followed the surgeon's hands, so hundreds of medical students could watch closely. RCA developed radar on-board commercial passenger planes, enabling those planes to search ahead for storms.

That culture led to thousands of innovations that shape how we all live every day. RCA lost some of its culture when General Sarnoff left, but today the company is reinvigorating it—a return signalled by the reappearance of Nipper as the company logo.

The Devastation of Unified Corporate Culture

Emerging new cultures and subcultures in business and industry—the personal computer culture, the robot culture, the CAD-CAM culture, the regional and state culture—all bring with them different attitudes, communication networks, symbols, agreements, and prejudices. As industries and markets become more sophisticated, specialized, and segmented, more subcultures will assert themselves—the product of the third derivative pace of change and multivariate forces working on today's manager.

These changes are challenging corporate cultures. The erratic nature of cultural shifts in a company and the difficulty managers, executives, and workers have in seeing cultures beyond the one they work within cause the company's culture to become fragmented; it becomes harder to reach a consensus. A set of barriers grows up against further change, especially against changes sought by management to stimulate innovation.

Successful managers understand this devastation and take steps not only to prevent it, but to employ the turmoil and commotion to innovate change. The trick is for managers and executives to understand the complex relationships that exist between innovation and corporate cultures.

Innovating Change in Corporate Culture

Gifford Pinchot, III, writes in his book *Intrapreneuring** that the first step in encouraging innovation in a corporation is to introduce new forms of managing. Pinchot is hardly alone; nearly every "excellence" book has suggested that the key to creating the factory of the office of the future is to break down old

*Gifford Pinchot, III, *Intrapreneuring* (New York: Harper & Row, 1985), p. 36.

hierarchical organizations and replace them with new ways of doing things.

This is good advice but extremely difficult to follow. Most companies cannot even get to the question of how to encourage innovation because they are stuck on the problem of dealing with employees who resist change. The key to innovating change is not a "skunkworks" in a spare warehouse or a tolerance for "intrapreneurship," but a corporate culture dedicated to dealing with change. The problem is not how to change the product; the problem is how to change the corporate culture so that the employees will accept new ideas, methods, products, processes, and understandings of what the company is and what it does. Unless managers know how to bring this change about, intrapreneuring, skunkworks, managing by wandering around, and all of the other innovative business principles will do a company about as much good as fad diets do for its executives.

Ten Steps to Change Your Company's Culture

Changing your corporate culture doesn't require magic. What it does require is down-to-earth action that will set a good example at the top, show some sensitivity about the cultures that are resisting, and unify the company behind a common set of attitudes. The following ten steps will help you in your effort to create a culture supportive of change.

1. *Start at the top.* If the top executives are not committed to change, no one else will be either. Companies that have become innovative have CEOs who are champions and sponsors of change. David Sarnoff challenged all of RCA's employees to stretch themselves through innova-

tion constantly, creating over time an innovative company.

2. *Use signals to alert your managers.* Some managers will take to the cultural change and some won't. Avoid the fight by initiating a self-selection process through the use of signals. The managers who are attuned to a culture of innovation will read them correctly. Those who are not will become victims, realizing only too late that they did not move with the changes.

T. Mitchell Ford, Emhart Corporation's CEO, published an article in *Financier* magazine* announcing that one of his goals was to raise Emhart's technical I.Q. Later the board of directors created a Technology Board Committee to signal the intention to support new technology.

3. *Correct or remove the hidden obstacles to innovation.* Many manager incentive compensation plans are based on short-term performance rather than long-term strategy; this is often a barrier to long-term innovations. Removing such barriers requires a shift in corporate policies to encourage innovation by altering the company's incentive compensation plan so it rewards long-term development as well as short-term results. Emhart and Sears used this technique to redirect management's emphasis to include not only short-term operating improvements but long-term strategies as well.

4. *Create a unified corporate culture.* Trying to innovate in a company that has more than one plant or division can be impossible if independent units remain separate in their thinking, especially if the products depend on integrated operations.

*T. Mitchell Ford, "Patron, Sponsor and Supporter of Raising the Technological IQ," *Financier* 3 (June 1984): 44–46.

By contrast, the executive who is able to forge a unified corporate culture is likely to create a successful company. Emhart was able to pull together the diverse divisions—Kwikset, Bostik, Pop Rivet, United Shoe Machinery, Mallory, and others—by stressing the competitive nature of the markets they were in. The result is a diverse firm with a market lead in virtually every industry it operates in.

5. *Create a marketing culture, not a sales culture.* Addressograph-Multigraph failed because of the nature of its sales culture; the company was only interested in selling, not in understanding its real customers' needs. By contrast, Ray Koontz at Diebold purposely created a marketing culture so that everyone in his company, even the sales reps, saw the opportunity for innovations that revolutionized the banking industry. The lesson is to tune the company's *culture* to the customer's needs, not the company's *products.*

6. *Create a listening environment.* Diebold's Koontz also demanded that his people learn the art of listening to their customers. His management team would meet with bankers during an American Bankers Association convention, and their rule was to listen to the customers, not to talk to each other. Koontz would break up any congregations of his managers and send them back to the customers to listen.

This constant emphasis on listening produced sales and service forces that constantly brought news into the engineering department about new developments in the field and refinements mentioned by customers, news they had obtained simply by listening. Similarly, managers and executives listened carefully to what they were being told by field people and consistently beat their competition with products that were quantum leaps in the industry.

7. *Don't marry one technology.* Even if your company succeeds by using a particular technology, always continue some experimentation in competing technologies and make sure all engineering and design people are involved in it. Also, find people with experience in technologies other than the ones your firm works in and rotate them through your research operations and brainstorming sessions. Otherwise, your company will become blind to competition from new technologies.

During the early development stages of the transistor, germanium appeared to have the edge in cost and performance over silicon, and several companies (including RCA) believed that germanium would ultimately be the transistor material for the solid state entertainment components. When silicon emerged as the leading material, companies that were married to germanium could not assemble the talent and experience needed to switch materials and had to drop out of the transistor market. But companies that had experimented with both materials and developed a *transistor* culture familiar with both materials profited from the transistor boom.

DON'T MARRY ONE TECHNOLOGY: THE STORY OF GADOLINIUM SELENIDE

During my tenure at Nuclear Corporation of America, our company's Research Chemicals Division announced that it had found tremendous energy-producing properties in a rare earth material called gadolinium selenide. Research for the Navy revealed that the substances had properties so great that it could generate power fifty times greater than any other material. Since Research Chemicals was the largest U.S. supplier of rare earth metals at that time, the announcement created a stir in the scientific community.

There was great interest in the material, and therefore I quickly decided to visit the Research Chemicals labs in Burbank, California. I found a strong culture oriented toward rare earth technology. As the new executive vice president visiting for the first time, I was treated to demonstrations, presentations, and projections of what would evolve from the significant breakthrough in gadolinium selenide research. After many hours of discussion, I finally got to see the magic material.

What I discovered was that the researchers could provide very little in the way of independent confirmation of the laboratory results. When I suggested to the scientists that an independent viewpoint would be desirable, my suggestion was not well received by the lab's scientists, who believed themselves to be the leading experts in the field. They told me that any outsider we might summon would not understand this particular emerging technology.

I could not accept the notion that there were no experts on the outside, so I demanded that we find someone qualified to examine the work of our scientists and determine the validity of the findings.

After conducting several technical audits, I discovered that the researchers had used a common laboratory practice that must be guarded against in any emerging technology. The materials being used were prepared under laboratory conditions in order to produce coefficients that could result in a theoretical prediction of its thermal-electric potential. The scientists discovered that one piece of the material had optimum thermal-electric coefficients, whereas another piece, while it did not have the same optimum coefficients, did have other properties that were needed as well. They then hypothesized that *if both of these properties could be obtained in one piece,* the resultant material would contain the high thermal-electric efficiencies that were projected.

But these two properties were not contained in a single material; they were in two materials that did not appear to be compatible. The "if" was pure conjecture and far too flimsy a basis for publicizing a breakthrough.

This conjectural technical approach is not unusual during early stages of experimentation. Many laboratory experiments start this way: A particular experiment reveals a potential of a good result, a secondary experiment reveals another potentially encouraging result, and the researcher tries to combine these experiments under optimum conditions to achieve a desired goal.

What was wrong at the Research Chemical labs was that the organization's technical culture was so strong that the scientists saw no need to confirm their theoretical results in practice. They felt that confirmation in some actual field test was not of major significance. Hence, they publicized their results long before field tests were conducted and long before the data should have been presented to technical societies. Had they reported these findings as an interesting laboratory observation, perhaps such premature announcements might have been justified. The results, however, were presented as a experimental finding that would lead to a significant breakthrough in the generation of energy. The scientists were wedded to the technology and would broach no intruders.

It was in this laboratory that I formed an adage that has remained with me ever since: "Beware of companies who are strongly married to a technology." The scientists involved, who were also managers of their company, truly believed that the rare earths had these properties, whether or not it was really true. And so they saw no need for independent confirmation.

A company whose business is tied to an emerging technology must be scrupulously objective. The managers must understand the limits of each technology and must evaluate each turn of events. When in doubt, experts should be called in who can substantiate the results or evaluate the methodology of the experts within your company. Until an emerging technology can be moved from the laboratory into a demonstration of feasibility for the potential commercial applications, managers must review the result as simply an R & D laboratory demonstration.

8. *Don't let product innovations fool you.* A single product innovation that becomes successful can mask the fact that a particular market is in decline because your market share is increasing. In fact, this is exactly what happened at Singer when the Athena 2000 became such a big success. One faction within the company believed that the entire home sewing market was on the rebound and saw no need to move other innovations on the Singer drawing board at the time.

9. *When you change cultures, retain the innovators.* The RCA corporation of the 1950s operated like clockwork

when it came to achieving technological innovations with electronics. But when the electronics/innovation culture that had fostered color television was replaced by a culture dedicated to administration, RCA lost much of its power as a leading innovator—many of the innovators departed, or they were reassigned to areas where innovation did not have the same emphasis that it had during the color television era.

10. *Always be ready to change cultures.* When Nucor could not keep up with changes in electronics, it moved into specialty steel. When Singer realized the downturn in sewing was permanent, it moved into electronics. When Emhart saw the threat to glass bottles from plastic bottles, it entered the packaging machinery business, manufacturing and marketing packaging machinery for new package materials.

No matter what culture your firm has settled into profitably and comfortably, change will eventually come along and jerk your business into a new environment. Be ready for it by constantly scanning the market and the society for trends, opportunities, and innovations, and then don't be afraid to use them when the time comes.

RIGHT CULTURE, WRONG COMPANY: THE NUCOR STORY

Nuclear Corporation of America, now known as the Nucor Corporation, was a small firm put together by acquiring entrepreneurial organizations in the late 1950s. Nucor owned a vacuum tube division known as Central Electronics, which I unsuccessfully tried to shift from vacuum tube technology to transistors. The people at Central Electronics were stuck in a vacuum tube subculture that was part of a larger culture made up of electronic engineers, ham radio operators, and electron tube inventors.

This culture firmly believed that its electronic tubes were irreplaceable. They reinforced this belief by pointing to the heat environmental problems the early transistors were having at the time as evidence of tube superiority. I saw it differently, and as executive vice president, I was determined to get Nucor into solid state as quickly as possible. But the tube culture would not have it. The R & D organization of Central Electronics would not shift its culture to working in solid state, and it took every opportunity to chronicle the problems of transistors and block my attempts to bring transistors into our research programs. I finally had to sell Central Electronics to Dumont Corporation, which stayed with tubes.

Nucor was never able to develop a solid-state electronics division. Its overall corporate culture had been devastated by the tube culture of the Central Electronics Division. Although its management team wanted to innovate and sponsors for the innovation included the CEO and senior officers, a technological culture within the company delayed the move long enough to thwart it.

I failed to change the culture of Nucor because I did not recognize the effect that subcultures can have in resisting change. Moreover, I committed a common mistake made by new executives: I tried to impart at Nucor the culture I learned at RCA. As a consequence of the Nucor experience, I learned an important lesson: *What works as a culture at one company will not necessarily work at another company.*

I tried to treat Nucor as if it were RCA, but it wasn't: RCA was a unified company with an overriding, pervasive culture, while Nucor was an amalgam of individual subcultures. At the time, I did not understand the difference between the two. With no overriding corporate culture, the individual subcultures fought hard against any changes aimed at them specifically; the subculture was their only identity.

I should have tried to lay an innovation-oriented corporate culture over the technologically based subcultures. Creating a company-wide culture is much less threatening because it does not single out any group of people for special treatment. In a unified company like RCA, there was no need to try to change individual subcultures because they were all tied together by RCA's corporate culture. But Nucor has no such binding force. One was needed before individual culture changes could take place.

Ken Iverson, who succeeded me at Nucor, was able to achieve this very idea—not by going high-tech but by going low-tech. He instituted a new corporate culture aimed at innovative methods of making steel and then divested the company of those high-tech units that would not adapt to this new culture. Today Nucor remains one of the few profitable U.S. steel companies.

The lesson of this experience is that a new executive, or any executive faced with the task of creating an innovative firm from a collection of different units, should not try to dominate subcultures from an isolated viewpoint but should bring aboard a management team that can develop an innovative and workable corporate culture. Part of this strategy may require the company to divest itself of units that can't accept an innovative corporate culture, even if they are profitable and otherwise an asset to the company. In the long run they will become a liability.

PART III

Innovate Your Business

7

3-D MANAGEMENT

When I worked at RCA, I developed a number of programs that were funded jointly by our corporation and our customers to examine potential products for a particular market area. One such program involved RCA and the R. R. Donnelley Company, a large Chicago printer and publisher, in exploring the possible application of electronics to printing and publishing. These programs brought RCA engineers and marketing personnel together with Donnelley personnel on a routine, day-to-day working basis. Both companies achieved a better understanding of the applications of electronics to printing and publishing as well as a better understanding of each other's top management attitude toward the program.

When I was vice president in charge of research at Singer, the R & D department funded Singer's service organization for the development of a training simulator that would help train service personnel on the Athena 2000, Singer's electronic sewing machine.

Through R & D funding of this project, managers within both R & D and service worked closely together. The service engineers had to actually construct the simulator. Thus, both R & D and service had an interest in the success of this program.

At Emhart, Bob Douglas, the vice president of R & D for the glass container machinery group, joined with six customers in a venture to learn how to strengthen and lighten glass as a future container material.

All these examples have one thing in common: They ignored the communication paths and boundaries laid out by the traditional organization chart. They recognized that a company is larger than its employees and its plants and that innovative processes require more than communication along hierarchical paths. In other words, they understood a company as *three dimensional,* and they used that understanding to make innovation profitable. I believe that in the future, no manager will be able to operate a successful company or unit without understanding this sort of 3-D management.

The Curse of Boxes and Solid Lines

One of the biggest problems with business today is the organizational chart. Derived from classical organizational theory, organizational charts are one- and two-dimensional entities that break up the company into formally defined functions (boxes) and lay out rigid pathways (solid lines) along which communication between these boxes takes place.

Once the boxes are filled by people performing specific functions, almost everything else follows logically from the chart, according to the theory. Management becomes a matter of putting the boxes to work solving problems. And planning is simply a matter of understanding the organization as a closed system: information flows up, decisions flow down. The organizational chart is altered slightly over time as boxes are added and subtracted, but the basic, rigid concept never changes. Classical theory assumes that the components of an organization are rational, efficient, predictable, effective, and understandable. This is the basis of the traditional organizational charts with which we are all familiar.

But companies are not one-dimensional. Information doesn't merely flow up or down the chart; it flows all around the company through informal networks, personal contacts, dotted-line responsibilities, matrix management, program management, and other similar concepts. And even if functions stay inside one box, the people performing them don't. They move around the organizational chart all the time—not just when they change positions but in formal and informal groups that constantly spring up and change for various projects and purposes.

Furthermore, the boxes and solid lines fail to account for the most important change taking place with present and future managers. Many of the most vital elements are not inside the company at all. Indeed, companies can no longer think of themselves as entities bordered by the factory gate. They are becoming intertwined with universities, customers, suppliers, other domestic and global companies, communities, and government agencies. The organizational boundaries are not boundaries at all; they're more like porous membranes through which a two-way flow of people, ideas, and resources is a constant reality.

As a result, the rigid, one- and two-dimensional organizational chart is rapidly becoming obsolete. A new approach must be adopted, one that recognizes that management today requires not a chart but a dynamic, organic, three-dimensional interaction that encourages creativity while it systematically directs resources. We can call this approach 3-D management. In laying out the principles of 3-D management, we emphasize not just functional relationships but human relations as well as the total environment and the sometimes amorphous interrelatedness within it that creates a third dimension totally ignored by classical organizational theory.

In the 3-D approach, managers must think in comprehensive terms about the nature of their organization and their roles in it. If a problem arises in one area, it is not up to one functional unit, one "box" on the organization chart, to solve it; rather, the entire system focuses on the problem. The manager's job is to link functions together with the environment so that problems can be solved and programs developed and implemented on a collective basis.

For example, when the Director of Currency was considering legislation restricting branch banking in the United States, Diebold's managers assigned not only the legal department to deal with it. Since R & D and marketing were just as concerned about the potential effect on the company's business, they joined forces with the legal department to achieve a better understanding of how the branch banking legislation might affect all departments and devised methods to mitigate that impact. This multidimensional approach led to many of Diebold's branch banking and drive-in products.

As with other lessons I have tried to convey in this book, the underlying lesson of 3-D management is this: Business managers must learn to move from a world of stability into a world of change. The innovative structures they will develop through 3-D management will not necessarily be permanent. They will have to change in order to meet current conditions in the five critical domains of change. But managers will have to adjust to one constant principle of 3-D management: they will have to learn to work *routinely* with the total environment that surrounds and supports the company including consumers, universities, government, global issues, finances, and many people who have been previously considered outside of the scope of the company and its managers.

How to Think 3-D

3-D management requires two basic changes in thinking. First, managers will have to see and work within their company in three dimensions. And second, they will have to modify their consideration of timing and the rate and speed of change.

Seeing the company in three dimensions. 3-D thinking begins with an understanding of the three dimensions of a manager's work world.

The first dimension is made up of the people who work in the manager's direct area of responsibility. It includes staff and line people who report to the manager and assistants and advisors who are part of the manager's unit or department. These people constitute the first line of communication for a manager, the people who implement the manager's ideas by applying their own. They appear on the organizational chart in the major functional box directed by the manager and are connected by the solid lines that signify "report to."

The second dimension includes people in the firm with whom the manager communicates constantly, but who are not within the manager's direct area of responsibility and who do not report to him or her. The manager needs these people to achieve objectives, but their relationship is based on communication, negotiation, and cooperation rather than on authority.

The third dimension, the one not traditionally considered in organization theory, consists of people and organizations in the five critical domains of change: technology, worker and customer values, demographics, the political and legislative environment, and global economics. This dimension involves the manager in literally hundreds of relationships—some direct and long term, some indirect and short term—with people and organizations outside the company. 3-D thinking involves the constant integration of this dimension with the two dimensions inside the company on a day-to-day basis in a manner that is profitable.

All managers work in the first dimension as part of their daily routine. All managers work in the second dimension in an informal but usually continual process. Some companies have formalized the second dimension through the use of ad hoc task forces or even standing committees to coordinate the programs of several managers. Program management is a successful second dimension concept. Few companies work routinely in the third dimension because it is foreign to the traditional concept of what a company is—the concept that is taught in business schools and embedded in the view of managers on a one-dimensional organizational chart.

But all of these dimensions actually blend together in the reality of the company. For success in the future, managers must systematize and formalize this blending to take advantage of all the opportunities and resources available to him or her and to see and anticipate the full spectrum of changes the company faces.

Expanding 3-D thinking to encompass the fourth dimension of time. There is another dimension that concerns all managers: time. 3-D management requires a manager to pay even closer attention to time because, as the number of variables increases, so will the rate of change. In particular, time is affecting management more than ever before. In a variety of ways—decreased development periods for bringing new products to market, increased speed of communications and transportation—time is, in effect, speeding up for managers.

Thus, managers must monitor and act quickly in ways that will involve them in the five critical domains of change. No longer can they take as much time to accomplish objectives as they once did. 3-D management requires them to compress time by speeding up everything from their decisionmaking to the lag time between customer orders and final delivery. Product development cycles must be cut by at least 40 to 50 percent. What formerly took days and even weeks must now be compressed to minutes and hours.

3-D Organization

The limitations of the traditional organizational chart have spawned one concession to the complexity of real-world relationships: the dotted line. Some boxes that are not connected with solid lines but exist on the same level in different parts of the organization are connected with dotted lines instead. For example, on most organizational charts, a corporate R & D director and a divisional director of R & D will be connected to each other by dotted lines. Thus, if you ask most corporate staff managers to explain their relationship with counterparts throughout the corporation, they will say they have dotted-line responsibility.

But this is really just an oversimplification that allows management to avoid having to explain relationships that don't fit into the traditional organizational chart. Dotted-line responsibility can mean anything—weekly meetings on a standing task force, sending copies of reports and memos, a meeting in the corridor—or it could actually involve both organizations in accomplishing key tasks in a structured, formal manner.

These dotted lines are meant to show that everybody is working together, cooperating and exchanging information and generally getting along. Yet, we all know this is rarely the case. For example, in his book, *Iacocca,* Lee Iacocca tells how he was stunned to discover that dotted-line relationships simply were not functioning at Chrysler when he took over:

> *What I found at Chrysler were thirty-five vice-presidents, each with his own turf. There was no real committee setup, no cement in the organizational chart, no system of meetings to get people talking to each other. I couldn't believe, for example, that the guy running the engineering department wasn't in constant touch with his counterpart in manufacturing. . . . Nobody at Chrysler seemed to understand that interaction among the different functions in a company is absolutely critical.**

Clearly, dotted-line relationships are valuable as sources of interdisciplinary cooperation and communication inside a company. And yet traditional organizational charts cannot categorize or even depict them with precision. Unless these specific functions work together very closely and systematically, the company will exist in permanent disarray, if not chaos.

In particular, traditional organizational charts, even those with dotted-line relationships, have failed to deal with one of the biggest problems in running a large corporation: communication between corporate staff and op-

*Lee Iacocca, *Iacocca* (New York: Bantam Books, 1984), pp. 152–53.

erating units. Method after method for organizing this relationship has come into vogue and then faded away, leaving both sides confused and misinformed.

For example, I can recall cases in which the corporate staff functioned as a monitoring organization to oversee progress of the major projects in the operating units. This structure proved cumbersome for both corporate and operating units. It required many written reports, most of which were produced too late for corporate staff to provide any assistance or corrective action.

Thus, I believe, the dotted-line concept needs to be replaced since the lines merely depict compromises rather than resolve what truly needs to happen. A manager's responsibilities must increase beyond their traditional boundaries, and these responsibilities must be explicit, systematic, and well understood. To paraphrase Iacocca, the person running the engineering department must be in constant communication with any number of people, not just a counterpart in manufacturing but also with a counterpart in several universities and customer organizations. In the future, managers will be judged by their ability to interrelate with people in other functions throughout the company and by how well they create innovative structures within the corporation and with external organizations.

Cubing the Company

The first step toward managing the company in three dimensions is to reconceptualize the way we bring our managers together. For example, at Emhart we cut across corporate and divisional boundaries by forming task forces. We visualized these task forces as operating as if the specif-

Figure 7–1. 3-D Task Force Concept.

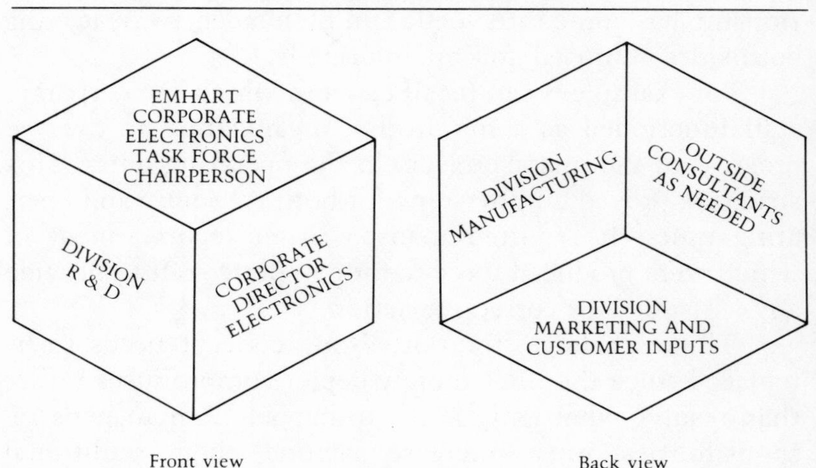

ic functions were the surfaces of a cube, similar to Figure 7–1. Each surface represents a function in the firm such as marketing, R & D, or manufacturing as well as external sources as required.

When you view your firm this way, two things become immediately apparent. First, no function is superior because the cube can be rotated to put any function on top. And second, all functions interface along the planes that make up the sides of the cubes. Unlike a traditional organizational chart, this cube depicts how a task force like this actually works.

Depending on where a particular project is in the pipeline—in development, manufacturing, or marketing—a different function has the lead and thus is on top of the cube. By rotating the cube so that a given function is on top, you can see the relationships that are important to include in a task force that is concerned with that aspect of the project. (As I will explain later, such cross-boundary task forces are the key to 3-D management.)

I believe that computers will eventually permit us to use a more complex geometry to indicate the *relationships* that are important to the development of a successful innovation. When the cube is computerized so that it can be rotated in three dimensions and manipulated by a manager, these relationships can be highlighted. But for our purposes, it is enough to understand the concept that *the relationships that join functions are the key to creating and successfully marketing innovations* and that these relationships should guide the formation of cross-boundary task forces.

For example, if your firm is losing sales to overseas competitors whose products are not better but whose prices are significantly lower than yours, you can form a cross-boundary task force to examine the problem. You look at your company as a cube for a rough idea of *all the functions that affect the production costs of the product* regardless of where they fall in the organizational chart. You will find that besides the obvious ones of materials, labor, and shipping costs, the list also includes social benefits, inventory levels, parts delivery, and quality.

I discovered this in a review of manufacturing costs for many operating units with which I have been associated. By looking at all the dimensions of the process and forming a task force, we learned that the manufacturing raw materials inputs were very competitive and the labor costs were a small percentage of manufacturing costs, while, on the other hand, a host of overhead items constituted a major cost factor from one country to another. The task force we set up to examine this problem needed people from inside and outside our company who represented all of these areas including human resources, lawyers, and tax experts.

Cubing the company will be a conceptual exercise at

first, but it will help shift your thinking from the organizational chart to the three-dimensional reality of your company. If you are adventurous, you can use your computer to construct cubes or perhaps other 3-D graphical representations with the names of different functions and shift them to see interrelationships that bear on specific projects. Once you have seen this concept, you're ready to use it to set up cross-boundary task forces and illustrate to top management how your function will operate.

HOW TO MAKE COMPUTER GRAPHICS WORK IN 3-D MANAGEMENT

In searching for ways to convey the 3-D aspects of business management, I have found the computer to be a valuable tool. As we expand the capabilities of computer graphics, they will be able to supply us with three-dimensional visualizations, geometry, and concepts that we have been unable to realize as yet.

Computer graphics has been a tremendous help in mapping out the cross-boundary task forces at Emhart. One good example was a corporate-wide electronics task force. Permanent members included key electronics managers from each of the operating units, while the general managers and the marketing representatives from each unit were called in for specific meetings.

The mission of the electronics task force was quite simple: namely, to identify the emerging electronic technologies that could provide product innovation to satisfy customer needs throughout Emhart's operating units. An additional objective of the task force was to allocate seed money in partnership with funds from operating units whose products or processes would benefit from these emerging electronic technologies.

The structure of the electronics task force was represented previously by Figure 7–1, a cube with one side representing each of the operating units and their respective functional representation on the task force. At the first task force meetings, each divisional electronic manager identified the emerging technologies that might benefit their particular product lines. Once the task force had identified these emerging technologies, each operating unit's marketing manager and general manager reviewed the list, adding their comments as well as their own candidate technologies.

Initially, more than 100 applications of electronics technologies were identified for Emhart products. Significantly, most applications could be served by the base technologies depicted in Figure 7–2. It should be noted that along the other axis the cube depicts the organizations with which the task force would have to deal in order to develop demonstration models that would clearly illustrate how these electronic technologies would improve the position of Emhart's operating units.

Cross-Boundary Task Forces

At the heart of the 3-D management structure are the cross-boundary task forces. Cross-boundary task forces differ from the committees and task forces you are used to in three significant ways. First, they are organized with specific productive, profitable missions and goals. Second, they are given money and authority. And third, the mem-

Figure 7–2. Example of Emhart's 3-D Electronic Task Force.

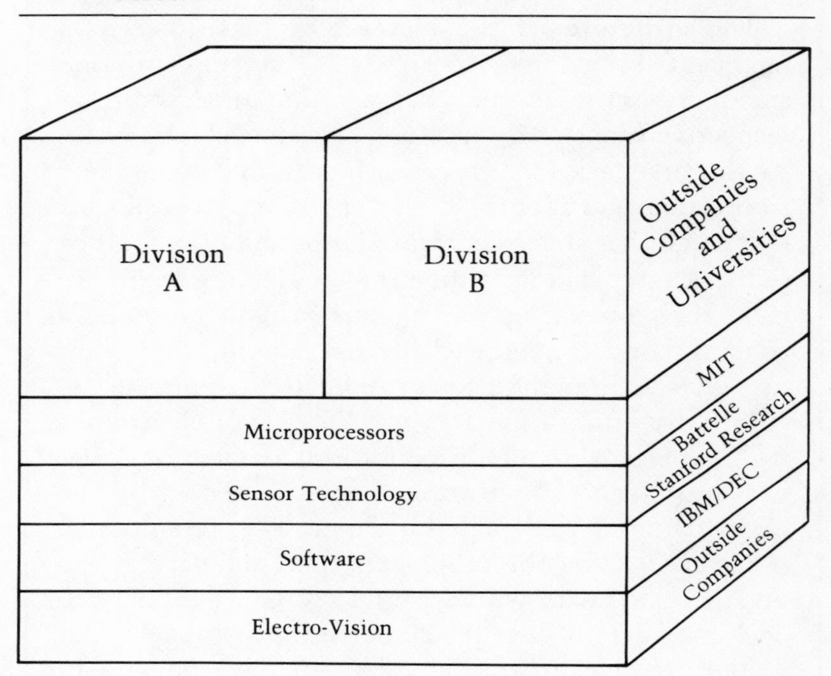

bership is based on functional utility, not rank or politics.

Cross-boundary task forces have no defined period of existence. They can be permanent, they can last through the life of their mission, or they can be formed to solve a specific problem. The length of time they operate is less important than the fact that their membership crosses both the company's internal and external boundaries.

Setting up cross-boundary task forces requires four steps: identifying missions requiring 3-D management, establishing strategic goals for the task force, establishing its structure and membership, and giving it money and authority to operate.

1. *Identify missions requiring 3-D management.* Not all missions require 3-D task forces. Some are contained completely within one function and can be approached by that function internally. The missions that *do* require cross-boundary task forces include those that are multifunctional in operation, that is, whose accomplishment requires resources or people from more than one function; those whose effects are multifunctional, that is, whose accomplishment will affect more than one function; and those whose operations relate to the environment outside the company or deal in multiple goals.

2. *Establish strategic goals for the task force.* The task forces are not merely advisory panels. They are operating units and must be handled like any other function in the company. Lay out goals and benchmarks and criteria to measure their progress and evaluate their work.

3. *Establish the structure and membership for the task force.* Each person with functional responsibility affecting or affected by the mission should be on the task force. Representatives from operations, manufacturing, sales, research and development, and others should be included as indicated by examining the cube that makes up this mission. The level of representation is usually from the managerial and executive level, but operating staff and other experts should join the committees when their insights and expertise are needed. Outside representatives such as suppliers and customers are also brought in when appropriate.

The task force leader is usually the member with the functional responsibility for accomplishing the mission, although more senior members of the company may be on the task force. The leader may be the person whose idea led to the innovation that the task force has been formed to implement, but, ideally, the leader should be a manager.

The senior people are frequently needed not for their leadership but for the resources they bring to the mission and the political clout they can supply throughout the company when an innovation needs a sponsor.

4. *Give them money and authority to operate.* The task forces are management units; they need to be able to spend money and hire experts. Each task force should have a budget and senior executives on board to expand the budget if needed. They need defined authority to spend money that allows them to bring in experts, give contracts to consulting firms, and borrow people and facilities from various parts of the company. In some cases, task forces will take over small parts of a manufacturing process to test innovations. This means the supervisors and managers whose output depends on that assembly line should be included on the task force.

The task forces need staff or access to corporate staff. Be careful that task forces in your firm do not fall into the trap that ensnares many corporate committees: becoming the home for staff people that other functions want to get rid of. Cross-boundary task forces will need the best people from the company.

Cross-boundary task forces are shaped to meet the goals of the mission. To give a sample of their range, here are a few examples from companies where I have used them:

- At Emhart, we established the electronics task force referred to earlier in this chapter to examine new applications for electronics to Emhart products. Members included electronics managers and marketing personnel from divisions within Emhart where we felt the potential for electronics applications existed. More than 100 potential applications were identified and 15 are now being actively pursued, with several ultimately resulting in products that were formerly mechanical and now incorporate electronics.

- Emhart's fastener group joined forces with Cincinnati Milicron's robotics division and the Ford Motor Company on a task force to develop automated stud welding systems for Ford.

- At Singer, a task force including both R & D people and manufacturing division employees worked to resolve the ten most unwanted problems plaguing the manufacturing division.

- An Emhart task force supported robotic workshops at Worcester Polytechnical Institute, at which manufacturing engineers, foremen, supervisors, and workers, could learn first-hand the application of robotics to the factory floor.

As you expand 3-D management in your company, you will find that cross-boundary task forces will increase in number and responsibility. Since they will not fit into the neat boxes in the organizational chart, another way will be needed to ensure that they are integrated into the company's operations and culture. A simple way to do this is to produce a book or catalogue of the task forces operating in the firm at any given time, either on paper or in a data base accessible throughout the firm. As task forces are formed and disbanded, the catalogue can be changed so that senior executives are kept fully informed of the management operations. At Emhart, a quarterly publication called The Emhart IQ (Information Quarterly) is circulated internally to provide to the operating units a task force update.

Cross-Boundary Thinking

The most important point about 3-D management is not that cross-boundary task forces should be established but that cross-boundary thinking should be fostered in all company structures. The task forces are simply a very useful tool. Once the executive and management culture of the company has grown accustomed to cross-boundary thinking, innovative 3-D management can operate continuously and independent of mission-specific task forces.

When Singer was developing new products, for example, the program manager at each step came from the activity that would have the responsibility for the next step. Thus, in the R & D stages, the program manager came from the product engineering organization. When the project moved to product design, a manufacturing engineer was moved in as program manager. When the program moved to the factory, a marketing representative took over. Similarly, at both Diebold and Singer, the design review team on major new product programs included representatives from manufacturing, marketing, service, finance, and purchasing.

At Emhart, we have also been able to move beyond cross-boundary task forces to cross-boundary thinking. For example, the corporate technology department purchased high-tech capital equipment that it believed would benefit operating divisions, even though those divisions did not understand initially how to use it to its fullest advantage. Once the value of the equipment at the operating unit was proven, captital funds were transferred so that the operating division acquired ownership of the equipment. Emhart's Dynapert Division acquired a sophisticated microprocessor video recording analysis system by this technique. When the equipment was transferred to Dynapert, the corporate function received a financial credit for the cash outlay.

Communicating in 3-D

3-D management is essentially a culture-shaping process in the company. A key part of 3-D management is communicating the concept. You must communicate more widely and more rapidly than most firms are geared to do. For example, don't write it if you can tell it; don't tell it if you can demonstrate it. Learn to develop quantum leap communication techniques for conveying to those in your managerial domain your objectives, strategies, and user needs.

These quantum leap strategies are not sophisticated or complicated. They can be as simple as conducting courses on how to translate user needs into product specifications, a technique that was used at Diebold and Singer. Emhart makes extensive use of videotape, satellite conferencing, reprints of key speeches, selected press releases, and internal publications to spread the word. All Emhart managers can take advantage of these communications vehicles to further their innovative potential.

At all three companies, top management reporting systems are established so that managers in charge of innovation can report at frequent intervals on the progress of their programs. The scheduling of these reports helps to focus the innovation and establish key milestone dates.

Scheduled presentations by managers to their peers, customers, and top management are also 3-D management tools. The presentations bring together people from different departments plus outside experts and customers. Assigning managers the task of making a presentation is an excellent way to give them better visibility, and it also forces them to understand their subjects better.

Innovating Change as a 3-D Manager

As a manager, your job is not only to bring about change but to gain the cooperation of the people within your 3-D domain who will be most affected by this change. This can be extremely difficult because, as I have suggested throughout this book, natural resistance to change is very strong. You must tread carefully.

It is important to recognize that even though you might understand why change is needed, your organization may not consider these reasons as problems in themselves. If you are the only person in your organization who sees that an area such as technology or worker relations is changing, be prepared to explain this development in terms of its total impact on the company, rather than as a specific reason. For example, if you believe your technology is becoming obsolete, don't argue obsolescence for its own sake; also argue that the customers will not buy your company's products anymore because they recognize the obsolescence as well.

You must prepare carefully when you present your change to the other members of the management team or to the people working under you. Make sure that you understand the effect of the change so that you can defend it. The employees under your direction will quickly recognize when you are not prepared for the change yourself.

It is also important not to be too ambitious when you make a change. Unrealistic targets will merely draw more resistance. So will attempts to institute too many changes at once, particularly if by making the changes you unwittingly create problems larger than those you are trying to correct. Even though you are looking for quantum leaps, it is better to proceed slowly at first, letting the members of your team gather the momentum, and then take bigger leaps once they are aboard.

You must pay particular attention to the way the people working under you react to your changes. No matter how beneficial a change might be to the organization as a whole, many people will decide whether they like it based on how it affects them personally. These people must be persuaded that the change will benefit them as well as the entire organization. Even some people who support and advocate your change may not accept it as readily as you might hope; most people are inherently threatened by anything except the status quo. Thus, you must take steps to bring these people around to your way of thinking. In order to innovate change, you must catalyze people.

THINKING 3-D: THE CASE OF THE CAD-CAM

Traditionally, the primary tool used in U.S. business management has been the departmental budget. It is a tool that grows logically out of the boxes-on-the-organizational-chart view of management: Each individual department owns a set of "assets" and budgets for these assets, justifying their budget request in functional terms. Money is rarely invested in developments or assets that can be shared by departments that perform different functions, even though different functions often must unite to accomplish the company's goals. Traditional capital expenditure procedures do not often permit the sharing of capital equipment.

Consider, for example, the case of CAD-CAM (computer-aided-design/computer-aided-manufacturing) equipment. Usually CAD-CAM systems are considered the property of the engineering unit. But the manufacturing department needs to understand and use the CAD-CAM equipment as well, at least if the company hopes to manufacture the products quickly and efficiently. And manufacturing can use CAD-CAM equipment to work with the engineers at the nitty-gritty level of design, when design modifications are needed to facilitate set-up and production. Indeed, CAD-CAM can form a rapid communication system between engineering and manufacturing, keeping them in constant electronic communication from design to final assembly.

Equally important, other corporate units can benefit greatly from access to CAD-CAM systems: service, sales, outside consultants and testing organizations that evaluate designs, and even the customers themselves. By limiting CAD-CAM equipment to the engineering department's turf, two-dimensional managers limit their firm's ability to move fast, modify quickly, and interrelate with all the people and institutions that affect the design and the manufacture of the products.

Several Emhart divisions share their CAD-CAM equipment. The engineering organization uses it to complete designs and drawings faster. But, equally important, it provides a solid modeling capability to the marketing organization. And sales and marketing people use CAD to examine what a product looks like before it is designed. Thus, their insights about the customer actually become part of the design process. Again, no department is an island; they all must work together for the company to succeed, and 3-D management give them channels through which they can communicate.

8

CULTIVATING INNOVATION AND CATALYZING PEOPLE

It is true that there is no one best way to innovate, just as there is no one best way to create a work of art; in the end, it boils down to the style of the management team and the company's culture. Even so, I have found that there is one truth in corporate innovation: The best managers don't manage in traditional ways. You don't really manage innovation as much as cultivate it. And you don't manage people as much as *catalyze* them.

When I say that innovation must be cultivated, I mean that managers cannot dictate and control it from above. They must encourage innovation by creating an environment in which it can grow and then carefully cultivate it so that it blossoms year after year.

And when I say that people must be catalyzed, I mean much the same thing. If managers encourage innovation in subtle ways, they cannot use traditional top-down management practices to direct personnel. For a company to succeed in cultivating innovation, people cannot merely take orders; they have to believe in their ideas, jobs, and companies enough to generate a constant stream of new ideas that can become successes in the marketplace.

Ultimately, the ability to cultivate innovation and catalyze people is, of course, a product of the corporate culture. If it is properly developed, as it has been at companies such as Diebold, this culture will remain in place and continue to generate innovations even as managers and executives come and go over time.

Cultivating Innovation

Managers who hope to cultivate innovation simply must be able to think about their work in the ways I have described in this book. They must view innovation as an ongoing process, not as a one-time event. They must understand the concept of the quantum leap. And they must understand the four techniques for innovation I laid out in Chapter two: ask the innovation question; market for the result, not the product; get to the root of the problem; and make innovation second nature. They must take a creative view of their jobs, companies, and lives.

It is easy to tell managers they should do all these things, but it is more difficult to suggest how. That is why I would like to pass along some of the most successful techniques for cultivating innovation that I have learned in more than thirty years in the corporate world. Of course, individual styles will differ, and no technique is guaranteed to succeed in every situation. And establishing structures can be a hollow exercise if the corporate culture is not responsive to innovation. Nevertheless, I have found that these techniques will help any manager to cultivate innovation in any company, no matter what their style.

One step toward creating a culture that delivers a constant stream of innovation is to set up formal structures that encourage and foster innovation. At RCA a formal New Business Development activity existed, cutting across corporate divisional boundaries in order to surface internal developments that could lead to new business opportunities for RCA.

At the Singer Company in the 1970s, an Office of Innovation was established for fostering innovative products throughout the company. This office established a mechanism for reviewing and accepting innovation proposals that had a future business potential for Singer. Through the corporate innovation structure, barriers that might prevent an innovation were removed. Innovative ideas were submitted from anywhere within the company. The corporate innovation office had a budget and a grant-giving structure so that innovative ideas could be funded, initially on a small scale, in order to bring the innovation along to a demonstrable stage when larger funding could be requested.

As a result of the RCA new Business Development activity, over seventy areas were identified internally within RCA as having potential for business exploration. Out of this activity emerged RCA's industrial electronic inspection business, which during the 1960s was a successful industrial business for RCA.

At the Singer Company well over 200 innovative ideas were submitted to the corporate Office of Innovation from which several submissions ultimately became programs leading toward commercialization. Singer's present line of handheld sewing and mending products was an outgrowth of this concerted innovation effort.

A structure for fostering and encouraging innovation can take many forms. At one end of the spectrum a formal organization such as that which existed at RCA and Singer was successful in encouraging innovation internally. At the other end of the spectrum a corporate officer either in technology, R & D, or marketing can institute a corporate-wide program for encouraging innovation.

Such programs can be designed to reward and highlight individual achievement in innovation. They can include mechanisms for accepting and reviewing innovation proposals, innovation budgets, and grant-giving structures; the removal of those barriers that might prevent the innovation process from being open to everyone in the company; awards, publicity, and other recognition for achievement in innovation; compensation systems to reward long-term development; and task forces with authority and money that are devoted to innovation task forces.

Once a formal program is up and running, then you can take additional steps to plant the seeds of innovation throughout your company. Here are eight steps that I find particularly useful. (The words management action or executive action following each task in parentheses suggest which level should undertake the task.)

1. *Examine and understand the needs of your customers.* (Management action.) In order to create a subculture that cultivates innovation, division managers must know what their customers' changing needs are. They must set up the systems and the links with customers to assess those needs and translate them into products and services quickly, efficiently, and reliably.

Motorola Company did this in 1984 when it joined with one of its key customers to develop and produce equipment that would reduce delays in communications between field personnel and home offices. The result was the Motorola KDT portable data communications system, which provides sales and service people in the field instant wireless access to home office computers, thus revolutionizing the way the customer's staff people use their time. It was an innovation resulting from a culture that asked, "What are my users' needs?"

2. *Create a culture that targets the situations where the user's needs are not being met—even before the user*

knows it. (Management action.) The gaps between a customer's needs and the way those needs are being met are the seedbeds of innovation. Managers who systematically identify those gaps with every customer on an ongoing basis will be able to cultivate those seedbeds rather than allow them to lie fallow.

Set up systems that routinely identify these gaps, customer by customer, product line by product line. Use the information as the basis for innovation research. Constantly examine what your customers do with your products and with competitors' products. Get to know their operations, problems, and future markets. If a manager encourages marketing, engineering, service, and salespeople to swap ideas, observations, and stories about customers, a stream of "If we did this, then they could do it better and cheaper" ideas from the manager's division will appear.

3. *Rank your customers' needs, the gaps between those needs, and the ways they're being met with your programs.* (Executive action.) Use correlation tests to correlate your development programs with the gaps. Are your programs aimed at filling those gaps? Are some of your budgets directed at introducing new products that have low correlation? Do your customers want reliability, while you focus your R & D investment on adding bells and whistles instead?

Not all your development need be aimed at specific customers or specific needs; some must be aimed at introducing new products to stimulate new markets. But the correlation between your customer's needs and your plans for new products should be high, or else you should have a good reason why it isn't. At the Singer Company I ranked customer perceptions of the value of sewing machine features. Using Spearman's Rank Correlation I correlated the size of the R & D budgets for feature developments with the customers' desires. Quite often the correlation was poor, signalling the need to rethink R & D efforts.

4. *Check with your user groups.* (Management action.) Don't rely only on yourself to see whether innovative ideas fill customer gaps. Ask the users themselves. Include them in the innovation review process. Incorporate their comments into the record of each idea. Give them a shot at improving the idea, suggesting how it should be delivered, serviced, configured. Wherever possible, joint customer development programs are highly beneficial.

5. *Test for quantum leaps.* (Executive and management.) In working with every innovative idea—whether it comes from R & D, the shop floor, or the customers—be sure to ask, "Is this the ultimate for this idea?" Probe deeply on this question, and don't let seemingly "practical" considerations limit your thinking. Ask the production people, sales team, marketing department, and customers for ways to take the idea even further or to accomplish its goals in a completely different and better way. For example, rather than develop new quality inspection tools, why not eliminate the quality problem in the first place.

6. *Set up internal grant-giving institutions to provide seed money to innovators.* (Executive action.) In Chapter 7 I explained the importance of giving each cross-boundary task force a budget as part of the company's R & D effort. But innovators need seed money to develop an idea initially so that they can demonstrate their need for additional funds. Establish a grant window inside the company to get these innovators started.

7. *Hold innovation presentations for managers.* (Executive and management.) Ideas need to be nurtured with further suggestions and synergistic applications. Managers whose units have come up with specific innovations should have the opportunity to present them to other

managers. A regularly scheduled managers' meeting to present innovations is probably the most efficient way to solve logistical problems. After a while, the managers will look forward to it.

These presentations should not be used to shoot down innovations but to build on them—to see how they might be used in other contexts and how other departments or divisions might handle them. Managers can also alter innovations by suggesting new uses and markets for them or by taking the design in directions that the innovators never thought of.

8. *Hold executive presentations for big money.* (Executive and management.) As innovations evolve, managers need to foster the support of top executives. Executive presentations can be regularly scheduled events or ad hoc, depending on the nature of the company and its logistics. The original innovator should be present, but he or she should find a manager or executive "champion" to actually present the idea and to explain what progress has been made and what action is needed to make it successful. These presentations give innovators and managers the opportunity to find a patron at the executive level who will fight for the innovation in budget meetings. Every successful innovation with which I was associated had an executive champion. Interestingly, in some cases that champion could be an executive from a customer's organization.

Catalyzing People

Cultivating innovation through the management techniques described above cannot succeed on its own. It must be coupled with a second strategy—catalyzing people so they will flourish in a freer, innovation-oriented environment.

Catalyzing people emphasizes training over directing, communicating over ordering, goals over quotas, and self-determination over routine. In part, the concept of catalyzing people rather than managing them is a response to a virtual revolution in one of the most important domains of change, worker values. But catalyzing people does more than simply respond to changing worker values. It also encourages innovation naturally. By giving employees more discretion and rewarding their successes, the process of catalyzing people multiplies the number of employee-generated suggestions and proposals that will improve products, the production process, and after-sale services.

At Singer and at Emhart, where we have had successful catalytic management programs, I developed several principles for catalyzing people and encouraging them to innovate on their own.

The first principle is that *excitement is the key to innovation.* Your excitement about a project will take your people faster and further than memos or staff meetings. When they see that a project is a priority with you because you enjoy it, they will enjoy its challenge and give the project their best. So you should generate enthusiasm among them through your own enthusiasm and your refusal to allow any opportunity or idea pass you by.

Just as important, however, is the idea that *ego trips get in the way of success.* If you are going to sustain the excitement, you cannot allow projects to become personal pets or obsessions. You must always make it clear that the project needs doing not to massage your ego, but because it would be fun and the accomplishment would be a notable one. This means that once other people in the organization have gotten excited about the project, you must step back and let them take the credit. Let them make the presentations to the task forces, talk with the CEO and senior executives, and get the awards and stories in the company newspaper.

There is one aspect of a project, however, from which you cannot step away—the problems. You must take the responsibility for failures or errors because *your people must have the freedom to fail.* If you inspire people to try things that are different and difficult and then let them get hurt if some of their attempts fail, you will surely kill the enthusiasm for innovation that you have worked so hard to create. You need to let innovators use every mistake as a teaching device so they will get back in the flow of work with renewed enthusiasm. They will only do so if they have the security of knowing that honest mistakes won't be held against them.

Another important principle for catalyzing people is that *all ideas should be listened to.* Some of the most preposterous suggestions have become major successes. If an idea is for a proposed product innovation, use the innovation maps described in the next chapter to obtain honest readings on its viability. When the results indicate that an idea is not viable, work with the employee to redirect the enthusiasm to other ideas or to eliminate the proposal's drawbacks.

If the idea addresses a problem in an assembly line, a service system, or some other ongoing operation, talk it over with others involved in the process, without identifying the source of the idea. If the idea does not have merit, go back to the originator and explain why this is so, and encourage him or her to go at the problem from another direction.

Finally, *it is very important to reward innovation in public.* Earlier I described the awards that Emhart gives for its top innovators and the publicity in company publications that encourages those who innovate. But let me add a word of caution: Make sure everyone gets the same amount of reward and recognition for the same level of involvement in innovations.

If a group of people are involved in developing a particular innovation, all of the workers should be rewarded equally. Unequal recognition will create hard feelings and kill the enthusiasm necessary to cultivate innovation and catalyze people. If the company is rewarding innovation throughout the firm generally, the rewards should be given only to those with clearly identifiable accomplishments. One California bank embarked on an "innovation and excellence" program in 1985 by designing a very attractive gold pin to be passed out to employees who met certain criteria for innovation and excellence. It was also given, however, to all senior executives as a matter of course, a move that angered many employees and diluted the effect of the program.

DON'T FORGET THE SHOP FLOOR: CATALYZING PEOPLE AT SINGER'S ANDERSON PLANT

In the early 1970s, Nick Mercadante, Singer's vice president of manufacturing, and I went to work retooling Singer's plant in Anderson, South Carolina, to manufacture the Athena 2000, the first electronic sewing machine. It taught me that no corporate innovator can ignore the importance of the culture on the shop floor in making innovations work.

We had to convert a mechanically tooled and organized factory to the production of electronic components. The work force had to be retrained and the management restructured. The purchasing department had to shift from buying metals to buying microchips. Quality assurance and testing had to shift from using mechanical tolerances to using some of the most demanding and precise tolerances and testing procedures in the industry. For everyone, the Athena required a major change in skills, attitudes, and concepts of quality.

The key to success was changing the shop floor culture from a mechanical orientation to an electronic one. The process was an ideal opportunity for me to experiment with productivity improvements through job restructuring. We were not bound by tradition, because the majority of the factory managers and foremen were new in their positions. I discovered that measurable improvements could be achieved in creating a new culture when management takes the time and interest to understand the causes of poor performance on the shop floor.

Here are some of the techniques I found effective in Anderson:

1. *Communication to and from the shop employees proved to be essential.* We structured formal communication links in both directions. All employees were informed of daily goals, problems encountered the previous day, and actions that were being taken to correct those problems. Employees were encouraged to submit their observations and, most important, every employee's submission received a management response.

2. *Top management walked the floor daily.* Nick and I walked through the plant at frequent intervals throughout the day. On these trips we observed and analyzed, but, most importantly, we listened to the

employees talk about their problems. This was more than just managing by wandering around. It was managing by systematically learning and analyzing every step of the manufacturing process with the employees. We wandered around, but it was the listening and the follow-through that were important.

3. *We replaced the "stretch targets" set by the executives and senior management with achievable short-term goals.* We found that the "stretch goals" necessary to cultivate innovation among executives and managers often had the opposite effect on factory employees. Confronting them with targets that were not reasonably attainable led many employees to give up. But setting incremental goals and moving the goals up step-by-step as the workforce progressed worked very well. We increased worker quality and output incrementally through competition; for example, contests to see which assembly line could produce 100 machines first, then 125, then 150. This generated enthusiasm and friendly competition but, more importantly, gave the employees an opportunity to set their own "stretch" goals.

4. *Quality checks were installed at each major station on the assembly line.* Machine operators performed their own quality checks. Quality Assurance staff audited the individual stations to assure that operator checks were being performed correctly, and Quality Assurance statistics were plugged back in to the work flow at each station. We learned that employees do not want to produce poor-quality products and that procedures can be established to encourage them to produce quality products by allowing them to check the quality of their own work.

5. *Assembly lines were credited only for good ma-*

chines accepted through final inspection. Defective machines were returned to repair stations adjacent to the location on the assembly line that produced the defect. At a glance, everyone in the factory could see where the defective machines were coming from, and manufacturing stations could easily see the results of their workmanship.

6. *Repair personnel became the trainers of assembly line personnel where defects originated.* Their rewards were based not on how many machines they repaired but, rather, on how well they trained their assembly line teams to reduce defects in the first place.

7. *Results for each station and each assembly line were prominently posted for all to see.* The number of good machines produced and passed through inspection were posted hour by hour for each line so workers could see their output and how they stacked up against each other.

8. *Changes on assembly lines were made only after the need for such a change was proven off-line.* One of the employees' biggest complaints was that management altered assembly line stations too quickly, actually worsening whatever situation we were trying to correct. After our experience with the first production lot, we discovered that the employees were right. As a result, we deliberately refrained from changing the assembly line until we were absolutely sure that the corrective action did solve the problem.

9. *Management recognized outstanding performance at every opportunity.* The assembly line team producing the first machine was photographed and received a special management ovation in the company cafeteria. Outstanding repair personnel became first-

line supervisors. And during our walks through the shop, Nick and I congratulated the employees who had achieved their goals or helped to improve the assembly process.

The results were staggering. Production went from 300 machines a week to 2,000. Quality was high, and the Athena's $1000 selling price was no barrier to record sales. One of the most gratifying results of innovation on the factory floor was the knowledge that even though the world's first electronic sewing machine was made in Anderson, South Carolina, it shipped to Japan for sale there.

3-D Catalyzing Sessions for Innovation

Everybody hates meetings—or so they say. But certain kinds of meetings can be used to stimulate a constant flow of innovative ideas. In fact, these 3-D catalyzing sessions are the single most important tool I use to cultivate innovation. These sessions, usually run in sets of three, are pegged to a specific need and involve employees, customers and outside experts. If run properly, they can be very successful in cultivating innovative thinking and catalyzing people to carry the innovations through to market.

I start with a need, a problem that needs to be solved or a challenge that needs to be met to stay ahead of the competition. I usually state the topic of the session in terms of a problem: "How can we solve this?" This stimulates a search for answers, helping to direct the sessions toward a useful outcome.

Session 1: Establish your 3-D group by bringing in people from the interested and affected parts of the company. Your selection of people to invite at this point will be mostly intuitive, based on the problem you want to solve or the user need you are trying to fill. I usually hold the first session in a conference room with a relaxed atmosphere. This also automatically limits the number of people involved to the number who comfortably fit around a conference room table, which is an easy number with which to work.

This first session has three goals: to explain the need or the problem to be solved; to communicate the source of information on the needs and establish sources for getting more information on it, including additional people at the next meeting; and to make assignments of contacts or research that must be done for the second session, to be held a week or ten days in the future.

It is important to include those who first articulated the problem or the need. Do so with introductory remarks such as "At the President's staff meeting yesterday the subject came up that there is a need for a particular new product," or, "The Vice President of Marketing said the other day that our products are too complex to operate." For example, at the Singer Company I kept hearing how complex sewing machines were to operate. This led to my establishing a 3-D group to hear of this complaint. Ultimately, these innovation sessions led to the successful commercialization of Singer's Athena 2000.

At this point, keep the discussion simple and general. If the topic is a user need that you have learned of through discussions with customers, present the situation to the group, stating who the customers are and what they have defined and their needs. Be sure to keep the first meeting focused on this user need. You are not looking for solutions at this point; you are simply planting the seed of innovation.

Be careful that you do not create an adversarial relationship by allowing your group to become critical of the person or organization who first identified the problem under discussion. Take care *not* to encourage people to defend their turf if it is being challenged.

You may have to be especially cautious if the topic involves the reliability of a product. Reliability complaints seem to generate very highly defensive reactions from managers and employees alike. As a manager you must permit the group to discuss these defensive responses yet still make sure that an adversarial relationship does not emerge. You can do so by pointing out that you yourself do not necessarily accept the validity of the reliability complaint but, rather, merely seek to discuss it. For example, sewing machine engineers at Singer constantly defended their designs against the complaint that sewing thread would jam. They were convinced that the sewers had threaded the machines improperly and resisted exploring more foolproof threading designs.

End the first session by making assignments. Ask specific people to meet personnel from the other areas where the same need has been identified. If the problem is with a particular customer, perhaps some of the individuals in the group should meet again with the customers. In all cases, people who attended the session should go directly to people in other parts of the company who are also affected, or to the customer, to discuss the problem.

If you have learned about the need from customers themselves, draw the marketing organization into the conversation before the catalyzing session. Thus, if members of your group feel that they need to know more about the customer's need, a marketing representative who understands the problem is already available to help. For example, it was marketing at Diebold that clarified the need for randomly generated safety deposit keys used in the Diebold safety deposit box system. Such systems greatly improves security by minimizing the discovery of key codes as in the case of a sequentially coded generating system.

Assign specific individuals to learn more about the user need and report back to the group at the next session.

Session 2: The purpose of this session is, first, to bring everyone up to speed, including new people in the group, through a briefing and reports from those who took assignments at Session 1; second, to establish links with outside organizations that might be helpful; and, third, to generate ideas. Thus, your role as organizer comes into sharper focus.

Because your objective is to generate enthusiasm and creativity, you should serve personally as the moderator. That way you will establish some objectivity and detachment from the different points of view. In fact, throughout all the sessions, remember to radiate enthusiasm, welcome all ideas, and assure the free flow of information, while, at the same time, discouraging negative comments. If some group members are not very vocal—and some members of every group will always fall into that category—make sure they leave with a thorough understanding of the needs and the approaches and meet with them privately to bring out their ideas.

Begin the session by reviewing your understanding of the user's need and the reports from those who have met with other groups. Have the members of the group identify the need in more detail. It may be necessary to bring in outside people such as consultants or the customers themselves to refine the group's understanding of the need.

Ideally, members of your group will begin to suggest who should be brought in to present more detail about the particular user's requirement.

All of these steps will help you learn who has the best perspective on the need. You may have to step in to do this if no one emerges from the group, but, in any event, you should soon have a good idea of who is best suited to solve the problem.

Generate ideas and give each one a hearing. Some will be discarded immediately, but others will grow and take on a life of their own. At this point, your role is to keep the creativity flowing and directed toward the need. Keep asking the innovation question. Whenever a solution looks like it might be the one, note it for further work and ask if there is an even better way it can be done.

At some point the session will run out of time or steam. Close the meeting by congratulating everyone for a good start and set a date no more than a week away for the final session. Encourage everyone to keep thinking about the ideas they have heard and to call you or others in the group if they have a flash of insight.

Session 3: In this session, ideas are selected for implementation, people come forward to take responsibility, and resources are allocated.

This session is action oriented; prior to the meeting you must establish with other executives that you will have the means to implement the first stages of any innovation that comes out of the meeting. This is important, because if you don't have top management backing after the third session, the effect of the meeting will be diluted, and you will lose any sense of urgency and enthusiasm you have generated. Also, if the group concluded, at the first two meetings, that two or more company organizations must work jointly on the issue, make sure you can deliver that cooperation before the third session. You might also want to consider inviting people who did not attend the first two meetings but have expertise that will be useful in bringing some of the ideas on the table to fruition.

Open the meeting by assuring the rest of the group of the resources and follow-up measures you have secured. Ask the best people in the group to summarize the progress to date, the ideas under consideration, and the distance that needs to be traveled to solve the problems at hand. Then moderate the rest of the meeting so that it hones in on the one or two ideas that hold the most promise for innovation and meeting the needs of the customer.

When the best ideas have solidified, end the meeting with a series of detailed assignments to groups that have naturally formed around specific ideas. Identify contacts in your office for each group with an assignment and set appointments to meet with them individually. You should wrap up the third session with one or more of the following accomplishments and actions:

- One or more innovation ideas has been selected for implementation.
- The necessary organizations within the company and outside of it have been put together as an operating team to develop the idea.
- Seed money has been distributed to conduct further experiments.
- Consultants have been identified who have the expertise to attack a particular portion or segment of the problem.
- A preliminary structure that will ultimately lead to a project has been put in place.

We are now ready to orchestrate the innovation to the marketplace.

9

ORCHESTRATION: GETTING YOUR INNOVATION TO THE MARKETPLACE

Let's assume that at this point, all your strategies have worked. The creativity of your innovators has brought forth a wonderful idea, a quantum leap that could take the company light years ahead of the competition. You have solved all the conceptual problems. The innovation is in some demonstrable form—drawings exist, customer acceptance looks promising, even a working prototype or a model is in operation. Now the time has come to move the product through production, sales, and service. To turn an innovation into commercialization, your company will have to risk big money in order to succeed in the marketplace.

This is the point at which most innovations fail—the point at which an innovation must be marketed and sold so that it becomes profitable. Despite this fact, the reasons for failure are not always clear.

While I was at Case Western Reserve University in the graduate school of business, I conducted a survey of 200 failed products to find out if there were patterns to why they never succeeded commercially. My survey included all types of products: consumer items such as the early combination washer-dryers, industrial items such as radio-frequency heat-treatment equipment, specialty items such as self-illuminating highway signs, and services such as irradiating food for preservation. All were innovative ideas; all had failed in the marketplace, even though they were successful technically.

Although I found close to 100 individual reasons for these failures, there were several common threads. The first was a poor identification of the users' needs and desires. Many of the unsuccessful innovators did not understand their customers' needs. (My finding was confirmed by Project SAPHO, conducted by the Science Policy Research Unit of the University of Sussex, U.K.,* a systematic attempt to evaluate the factors by which some inventions became commercially successful innovations while others failed.)

The second common thread I discovered was that the companies marketing them did not have a systematic process for discovering why products failed and then avoiding those problems. In subsequent discussions with the managers of these firms, I found that each firm *did* search for the reasons their particular invention failed, usually with a great deal of destructive finger-pointing: Marketing did not get the customers' needs right, manufacturing did not make it right, engineering didn't give manufacturing producible specs, and on and on.

Third, I found that those firms that had some type of process to *identify the causes* of product failure did not re-examine their findings over the life of their development programs. When factors affecting the market for their innovation changed, they failed to note or act upon the change during the product development stage. Similarly, when a technology changed, no one studied the change to

---

*Christopher Freeman, *The Economics of Industrial Innovation,* 2nd ed. (Cambridge, Mass.: The MIT Press, 1982), p. 113.

determine whether the product should also be changed. They ignored the five critical domains of change.

As the first step in devising a system to avoid product failures, I categorized these reasons for failure into six major patterns:

- *Market problems:* the product lagged behind competition; too much customer training was required to use it; and the product was introduced into the marketplace too late, when other innovations were coming in to replace it, or most significantly, the customer's culture was not ready for the product.
- *Marketing problems:* a lack of understanding about how much and what kind of promotion was needed for success, inadequate distribution channels, incomplete marketing specifications, and the fact that a product was not designed for the market at which it was aimed.
- *Technological problems:* a poor patent position; incorrect estimates of technical feasibility; and technical compromises made without checking with marketing to determine the impact on the customer.
- *Manufacturing problems:* new manufacturing processes required a learning curve longer than originally estimated; inability to deliver the product on schedule; use of new materials having unfamiliar characteristics; inability to meet or hold engineering specification; and failure to achieve quality levels.
- *Financial problems:* skipping field tests to conserve funds; reducing marketing and advertising expenses; cash-flow was less than anticipated, capital expenditures were higher than anticipated, or the total cost of commercialization was underestimated.

- *Other problems, depending on the type of product:* legal or regulatory problems, unanticipated employee training, purchasing problems arising from single-source vendors (or sometimes no vendors at all!), and inability to acquire the needed skills or type of labor force.

NOT ALL INNOVATIONS SUCCEED AS PLANNED: AN ENCOUNTER WITH WALT DISNEY

"Walt will see you now."

With this brief introduction, we entered the office of one of America's most famous creative geniuses. Oscars, Emmys, and other plaques and trophies lined the shelves in front of us. A sharp turn left past the shelves, then a quick right, and there was the great innovator himself, standing at his desk in a sweatshirt. Walt Disney did not function in formalities but in the dreams that shaped our culture.

I was there with David A. Thomas, an RCA vice president, to discuss the possibility of jointly making one of Disney's dreams a reality: "the City of Tomorrow."

Disney had been disappointed that he had not been able to buy more land around Disneyland in Anaheim, California, to prevent the unattractive strip development that now surrounded the park. The area of motels, souvenir shops, and other establishments around Disneyland did not reflect his vision. So he decided to build an entire city where people would live within the total Disney culture.

The proposal Disney had given RCA was a city of 65,000 people on 12,000 acres near Palm Beach, Florida. He wanted to use the city as a dramatic demonstration of progress in the art of urban living made possible by science and technology, particularly electronics. It would be both a functioning city and a tourist attraction.

We knew Disney was setting out to, quite literally, change the landscape of America, and we of RCA who worked on the project were thrilled with the innovative freedom in preparing a proposal for a joint RCA-Disney project. With Disney personnel under the direction of their technical director Ub Iwerks, we created dozens of concepts and working plans: electronically equipped and instrumentated hospitals, new methods of audiovisual education, automated and remote meter reading utility systems, totally automated and progammed home appliances, monorails, closed-circuit scanning systems for security and protection, technology to improve pedestrian and auto traffic safety, and improved city aesthetics through chemical horticulture and complete underground installation of all utilities.

Disney stressed that the project should be not only a technological development but a social experiment as well. He intended to call in the best city planners and universities to participate in creating a city that would be planned for the social living of the future. Clearly, he wanted to reshape American culture. Ironically, however, he ran into trouble because of RCA's own corporate culture.

Like Disney, RCA under General Sarnoff stressed innovation. But there was an important difference: Whatever RCA produced had to exhibit a potential payback in the marketplace to continue. The City of Tomorrow did not pass that test. RCA's cost estimates of the houses in the project came in so high that prospective home buyers would not have been able to afford them without subsidies. And even after they moved in, the subsidies would have to continue in order to support the technological marvels of the city's infrastructure. Worse, RCA's staff could see no adequate payback in the technologies that Disney wanted the corporation to develop. R & D costs were so high that they could not be recouped in a reasonable time period. For RCA, the City of Tomorrow would have been just too expensive, not economically justifiable.

Disney hoped the city could overcome its economic problems through tourism revenues. But here his own cultural myopia misled him. His idea was that the city's residents would be part of the attraction, always on view in their homes and on the job. Furthermore, he wanted the residential districts of the city to be divided by income. Many city planners and sociologists tried to point out that some people may not want their private lives to be on display to thousands of tourists, and, in any case, dividing residents by income was economically and sociologically impractical.

Disney also tried to persuade RCA to change some of its own corporate values. At the time, RCA was putting a lot of money into color television and computers, and General Sarnoff believed that undertaking something as large as the City of Tomorrow would be prohibitively expensive. But Disney suggested that RCA purchase real estate in and around the new city in hopes of profiting from land development. While the technology in the project might not show a payback, he reasoned, the real estate probably would. But Sarnoff thought the idea was too risky. Ultimately, the RCA executives supporting the program left RCA, stripping the City of Tomorrow program of its internal supporters and talent. Eventually Sarnoff dropped the program altogether.

Disney was unable to convince RCA of his *own* culture, which was to lead in innovation, using means such as the appreciation of real estate to make the idea profitable. At RCA, on the other hand, the business side dominated the culture. RCA's culture demanded that technological innovations be pursued for their own profitability, not as "loss leaders" that would enhance some other venture. Thus, the corporation was uncomfortable with the idea of undertaking a technology venture that had little payback potential in order to make money on real estate investments.

After Walt Disney's death, the Disney Company went on to redesign the City of Tomorrow into what is today EPCOT Center. Although EPCOT has been successful, it was not what Disney had originally envisioned. In the final analysis, Disney could not change what he helped to create: the culture of his customer—in this case, his potential joint venture partner and other financial barriers.

The Innovation Maps

To avoid these mistakes, I knew I needed a system that would do three things. First, it needed to keep me informed early on in a program about what failures could occur and why and to help me assess the probability of correcting those problems and continuing with the innovation successfully. Second, it needed to identify company partners for the introduction of new products. And third, it needed to bring those partners into the loop of responsibility so that they could correct the problems in their area, identify problems that could not be corrected, and apprise me of any changes that might affect the innovation's success.

The approach I developed brings all the managers responsible for commercializing an innovation into the process at the beginning of commercialization and ensures that their input is repeated and updated at key milestones during the process. It does so by using what I call "innovation maps" (Figures 9–1 through 9–6).

The maps can be used to obtain a quick "snapshot" of potential problems that innovation encounters, as well as a qualitative estimate of the probability that these problems can be overcome. The maps are meant as an information tool, not a bureaucratic control device. They are meant to bring people and groups together to commercialize an innovation, not to provide excuses to prevent innovation but rather to bring about corrective action. They can be used by anyone at any level and at any stage of the innovation. An entrepreneur (or an intrapreneur) can use the maps to gather support, identify potential problems and opportunities, and move the project along.

I have used these maps with virtually every product I have brought to market in the last twenty years. I created the first set of maps myself and have directed the preparation of other maps, but eventually other departments and divisions in firms where I worked began to design their own maps, using these as guides. The maps printed in this

Figure 9–1. Innovation Map: The Market.

Marketing Research Status	*Market/Customer Acceptance*	*Competitive Analysis*	*Product Life Cycle Analysis*	*User and Market Needs Analysis*
Recent marketing research—positive.	Positive acceptance projected	"Way Ahead" of competition.	Product at early stage.	Personal customer contacts identify user needs.
Positive indicators from credible sources.	Customer training required.	Superior features over similar products.	Product at growth portion of life cycle curve.	Needs analysis through user experience existing in company.
Updating required.	Other products initially need to be in place.	Competitive products have comparable features.	Product extends life cycle curve.	Market and user needs transmitted through sales organization.
Marketing research scheduled.	Major customer culture changes required.	Threat of new technology.	Further analysis needed.	Needs transmitted through other sources.
No plans to conduct marketing research.	Market/Customer acceptance data lacking.	Competitive analysis lacking.	Unable to position.	Incomplete or conflicting data.
Negative or uncertain results.	Resistance expected.	Indirect competitive threat.	Mature product. Mature market.	None.

Figure 9–2. Innovation Map: Marketing.

Marketing Specifications	*Market Development*	*Channels of Distribution*	*Promotion and Advertising*	*Service and Installation*
Specifications accepted.	Market knowledge exists in company.	Existing distribution organization.	Existing advertising and promotional programs.	Compatible with present organization.
Minor spec changes.	Market development information can be obtained.	Channels available within the company.	Minimal advertising and promotion required.	Retraining feasible.
To be issued.	Experience in existing markets could be applied.	New channels needed.	Not considered.	Major retraining required.
"Open" spec issues.	New product. New market.	Requires further assessment.	Promotion and advertising budget at a high level.	High installation and service costs projected.
Disagreement on "specs".	Market development information lacking.	Not considered.	Advertising budget unacceptable.	Unfamiliar installation and service.
Insufficient data.	Uncertain.	Optimum channels not available.	Advertising, promotion, and publicity, beyond capabilities.	Excessive warranty costs projected.

Figure 9–3. Innovation Map: Technology.

Project Status	*Schedule Feasibility*	*Patent Status*	*Product Features*	*Information Systems*
Fully designed. Field tests sucessful. Meets market requirements.	Projected 90% + to meet target completion dates. Product specs accepted.	Patent issued. Broad claims allowed.	Product has competitive advantage in existing markets.	Present systems adequate.
Feasibility demonstrated. Field tests being conducted.	80% projected. Minor spec changes.	Good chance of obtaining patent.	Expands existing markets.	No effect projected.
No field tests initiated.	60%-80% projected.	Competitive patent issued. Some coverage possible.	Features are merely "bells and whistles".	Additional capacity projected.
Design reviews identify potential major redesign areas.	Better than 50%. "Open" spec issues.	Licensing to be considered.	Product is a restyle or redesign.	New systems required.
Basic elements require considerable R&D.	Less than 50% of meeting schedule.	Infringement projected.	Other products exist with superior performance.	Requirements uncertain.
Technical feasibility uncertain.	No way to meet schedule. Insufficient spec data.	No protection possibile.	Easily copied.	Not considered.

Figure 9–4. Innovation Map: Manufacturing.

Producibility Analysis	*Facility Availability*	*Manufacturing or Process Technology*	*Quality and Reliability Analysis*	*Production Skill*
Designed for producibility. Standardization achieved throughout.	Facilities exist.	Technology now exists.	Product meets quality and reliability goals.	Now exists and available.
Designed for producibility. Departure from standards acceptable.	Existing facility needs minor modifications.	State of the art technology can be implemented.	Corrective action identifed.	Some retraining required.
Requires further analysis.	Subcontract or utilize other company facilities.	Technology identified. Appropriation initiated.	Quality and reliability assurance. Restructuring required.	Major retraining required.
Manufacturing has not "signed off".	Different type of facility required.	Technology being identified.	Targets not established.	Transfer of skills necessary.
Producibility reviews have not been conducted.	Seek external sources.	Technology has not been adequately considered.	Manufacturing quality and reliability assurance not adequate.	Skills not available.
Product has a poor producibility rating.	No capacity available internal or external.	Technology not available.	Product quality and reliability uncertain.	Total new technology and skills required.

Figure 9–5. Innovation Map: Financial.

Sales Potential	*Profit Potential*	*Justification*	*Effect on Assets*	*Financial Analysis*
Growth rate exceeds targets.	Profit growth exceeds targets.	Exceeds justification targets.	Achieves or exceeds ROA and ROCE targets.	Projected financial ratios above target.
Acceptable growth rate.	Profit levels acceptable.	Targets could be met through additional action.	Plant and equipment expansion within strategic goals.	Financial ratios below average but within strategic targets.
Below target.	Profit levels could be improved through cost reductions.	Further data required.	Added depreciation offsets profit potential.	Financial targets not established.
Sales levels uncertain.	Profit levels uncertain.	Inconclusive.	Needs further analysis.	Uncertain ratios.
Declining sales.	Projected profits below operating unit targets.	Critical factors could shift justification.	Unable to obtain required resources.	Key ratios below target.
Projected sales level unacceptable.	Unacceptable.	Justification far below "hurdle rate" targets.	Effect on assets below target ratios.	Financial ratios unacceptable.

Figure 9–6 Innovation Map: Other Functions.

Human Resources	*Training*	*Legal*	*Purchasing*	*Top Management Assessment*
Minor impact projected.	No training required.	No legal or regulatory issues.	Use present purchasing structures.	Favorable toward product and program.
Issues identified. Programs underway for resolution.	Minimal training required.	Regulatory, safety or environmental approvals obtainable.	Reliable vendor, multiple sources. Purchasing under control.	With resolution of open issues, program has management backing.
Major cultural change required.	External training support required.	Approvals required for use by customer or in manufacturing.	Long lead items or shortages on critical items.	Scheduling and project monitoring not acceptable.
Critical skills required. Compensation levels need review.	Major retraining required.	Unresolved legal issues.	Major components single sourced.	Dilution of management effort projected.
Open issues exist.	Training requirements not identified.	Market area or manufacturing highly susceptible to regulatory requirements.	Lack of suppliers.	Lack of management skills or major organizational restructure required.
Major problems foreseen.	Major training resources required.	Threat of sudden and damaging regulatory action or legal problems.	Unfamiliar purchasing.	No confidence in either meeting product performance, schedule, or both.

book are a result of constant refinement based on continuing examination of the reasons various innovations succeed or fail.

Setting Up an Innovation Review Board

But the maps are only part of the system I devised. The other part includes the project team and a review board.

The project team consists of a project manager and personnel required to bring the innovation to commercialization. The review board serves as an advisory group to help the project manager identify and obtain the resources needed to bring the innovation to market. The board also can provide the project with resources and help the program managers in their specific areas of responsibility. Board members can send signals to the organization that management is interested in getting this innovation out into the marketplace and can help chronicle the innovation throughout the firm.

The effectiveness of the board, of course, depends in large part on its membership. If the program managers have done their homework, powerful and responsible people can be recruited from the six critical areas where the potential for failure exists. For example, members might include other managers from marketing, service, financial, manufacturing, human resources, and top management. The seventh member of the team is the actual innovator, if that person is not also the program manager. If the program is expected to take several stages, and so will involve a "handoff" by the current program manager, the successor should be on the review board. But the actual innovator should always remain on the review board.

In this sort of situation, the position of project manager becomes highly sought after. Program managers can be virtually anyone within the firm, staff or line—engineering, marketing, or manufacturing people—but the post will put them in direct touch with senior management in the firm, giving them visibility and opportunity.

Circulation and Initials

To illustrate the use of the innovation maps, the project manager would initially assemble a briefing on the innovation for the people within the firm who will be involved in commercialization. This briefing should cover the innovation as well as estimates about its potential in the marketplace. A top manager's statement of support is a key part of this briefing.

At the end of this briefing, the maps are distributed to the participants, who are asked to study the innovation and return the maps by a given date, initialing one of the six boxes for every category. Initials in a solid box indicate that key players foresee problems with the innovation. White is, in effect, a green light, while grey is the equivalent of a yellow light, indicating to proceed, but with caution. The purpose of this procedure is to find out where the innovation stands with the key players who will have to implement it. No comments are requested at this stage, and no plan or program for implementing the innovation has been yet prepared.

Then the project manager reviews the returned maps for the initials in the solid areas and meets with those concerned to find out how to shift their initials out of those boxes into grey or white boxes. The team is called together to discuss the results of the meetings and consider recommendations about changing the solid initials. Some of these recommendations may be drastic: a new factory, perhaps a new distribution network. In some cases the project manager may disagree with the initials in solid areas and can appeal to the team for support to move them into grey or white areas.

The project manager works with the team to find strategies to overcome the problems or with a sponsor to prepare for a serious meeting with the key executive backing the project (in the event they are not the same person) to determine if top management will support strategies designed to overcome the solid initials.

If there is support, the key executive will give the project manager the green light to go ahead with the needed strategies. If the key executive cannot support the strategies needed to overcome the objections, the innovator and project manager must decide whether to abort the project, a decision that must not carry any indication of failure, or to seek other sponsors. They might also meet with the company president or the most senior sponsor and the person blocking the project to see if there is a way around the problem and how long that will take.

Selecting the Team, Getting the Budget, and Setting Milestones

The project manager then selects the team needed to solve the problems and implement the innovation. Together they set the milestones on the way to commercialization. The milestones that I believe should be used include:

- Deciding to commercialize the innovation
- Completing the program plan, manpower needs and budget

- Conducting design and manufacturing reviews
- Completing, testing, and evaluating the first prototype
- Designing the field test program
- Conducting the field test and completing analysis of results
- Reviewing manufacturing and marketing plans
- Procuring of tooling
- Evaluating initial production products
- Analyzing early customer experiences with product

During the entire process, the maps are used at the milestones and sometimes in between them. The program manager and the team must use the maps to follow changes, since the critical areas can improve but can also become worse. The changes are signified by initials, so the project manager and the team has a record of each key participant's evaluation of the innovation over time. Everyones's support or opposition is recorded on paper, and their reasons are explicitly stated. No one can say, "I didn't know."

At each milestone meeting, the individuals responsible for any delay in the program must report to the team and explain the reasons why. The program manager has overall responsibility for the innovation, but the system also identifies individuals who are responsible for problems as a way of supporting the program manager. The program manager can also bring exceptional performers to the team for congratulations.

I used this system with great success at Singer during the development of the Athena electronic sewing machine.

The project manager at the manufacturing stage, Ed Cherico, was not on my staff, nor was he a top manager; rather, he was a staff member who understood the process of innovation and wanted a challenge. When Ed was reviewing the maps, he saw that the vice president of manufacturing, his own boss, had initialed four solid boxes to indicate that Singer, which had a 100-year tradition of building mechanical products, did not know how to build an electronic sewing machine.

"So what?" Ed commented matter-of-factly. "That's why I asked for this job." And instead of killing the project because of Singer's mechanical culture, Ed and I went to Athena's innovation review board and devised a strategy to overcome Singer's shortcomings in electronics manufacturing. The group president, who was part of the team, gave us his enthusiastic support and considerable clout.

We assembled a management team from Singer's units that made electric motors and Craftsman power tools, all of whom had experience we needed in electromechanical assembly. We persuaded the South Carolina state government to set up schools and train workers in electronic assembly if we located the machine's assembly line in the state. We hired the number two quality-assurance manager from Singer's military assembly operations—Singer's toughest quality assurance unit and one that dealt regularly with electronics quality control.

Using the map highlighted the manufacturing problems immediately, even before drawings were off the board. Had we not used the map to pinpoint potential problems and start working on them early in the game, the project probably would have been cancelled as the manufacturing problems began to arise. As it was, the initials inched up out of the solid boxes as each milestone meeting showed that we had assembled more and more of the needed talent. When the signatures finally went into the white, we tooled up and produced the world's first electronic sewing machine.

MAPPING THE ATM AT DIEBOLD

The innovation maps were used at Diebold to bring the automated teller machine to market. As part of our first project meeting individuals initialed their maps. Most of the initials appeared in either the white or grey boxes. But a few of the initials appeared in the solid boxes and as a result the ATM hit several snags right off the bat.

At the time I first circulated these maps, I had worked at Diebold as a corporate vice president for two years. When we conceived the idea for the ATM, I decided to run with it. But I was dismayed that this project got red-flagged early in the game. Later, I learned that these red flags were probably the best thing that could have happened to the ATM project. Following the ATM through the innovation implementation process illustrates how you can use the maps to overcome barriers and solve problems on the way to the market.

The solid box initials were those of Earl Wearstler, vice president of the Bank Division. I was shocked. Not only had Earl expressed his enthusiasm and support for the idea, but we had actually worked together on development of the idea itself. And here were his initials in the solid box!

Earl's initials appeared on Figure 9–1 (the market), in the second column representing customer acceptance. He had initialed the box labeled "market customer acceptance data lacking," because he believed that we had no data on the reaction of bank customers to a machine doing what they were used to seeing live tellers do.

I just about flew from my office in engineering over to his office in the marketing building when I saw his initials. "Earl, why are you killing our ATM project?" I asked. "You wanted it to go just as badly as I did."

"I want it worse than you do," Earl replied, "but we have to recognize that banks are not all that forward-looking." Then he reminded me that even though customers have been standing in teller lines for decades, most banks had not even initiated simple innovations such as a queuing line.

As a result of that conversation, we developed the strategy of using a model ATM at the American Bankers Association meeting that year in San Francisco. That strategy gave us the green light to build a prototype. It was the hit of the show, and Earl's signature moved up to white. By using the map, we accelerated the prototype process, got an early field reaction, and developed strong support from the potential market, namely the bankers.

The victory at the ABA convention was key to overcoming another set of solid box initials that appeared early on Figure 9–6, those of Dwain Crawford, Diebold's corporate counsel. Dwain stopped me in my tracks by initialing the box marked "threat of sudden and damaging regulatory action." He was concerned that banking regulations might consider the ATM a branch bank and, therefore, subject to very strict branch regulations.

After seeing the prototype at the convention, the

American Bankers Association brought in an expert, Dale Reistadt, to attack this problem. Given the enthusiasm of the bankers at the convention, our potential customers wanted a positive answer as much as I did. Reistadt concluded that branch banking laws would not be a problem, and bankers were willing to take a risk on ATMs. As a result, Dwain's initials moved up to grey, reflecting his caution about the potential regulatory problems but removing a roadblock for us.

The third set of solid box initials belonged to Chet Chasin, my chief engineer, who flagged the box labeled "project status: basic elements require considerable R & D." On Figure 9–3 Chet believed we could build the machine we wanted, but he pointed out that we did not know how to read a magnetic card in a reliable and cost-effective fashion and that cash dispensers already on the market had an unprofitable habit of giving away too much money. He was also worried that customers would not want to learn how to work the machines, preferring tellers instead.

I worked out the solutions with him directly. We used a television tube to give the customer directions and assistance when needed. And to keep faulty currency feeders from bankrupting the banks, I suggested that we initially deliver money in packets, instead of counting out single bills. As for the problem of reading magnetic cards, we found that scientists from SRI and Booz-Allen were already working on this problem for other clients in other areas, and we hired them to help us as well. We also joined ABA's Personal Identification Project, which helped give the association and companies like ours insight into how ATMs would ultimately be activated.

As a result, we came up with a system that worked in the prototype, and two years later we solved the reliability problem. As time went on, the gathering white and grey initials produced a momentum for the project that the solid initials could not overcome. But, more importantly, the maps identified problem areas that helped me determine where to put my effort and my resources. The result, of course, was an innovation that has made Diebold the leader in a very profitable and rapidly growing market.

Afterword

Addressograph-Multigraph changed the buying culture of the United States by replacing the metal charge card with a plastic credit card that could be used universally. Diebold changed America's banking culture with the introduction of the ATM machine. Apple created a new culture by bringing sophisticated computers into the home, an innovation that, along with the telephone answering machine, unleashed the explosion of home offices and home-working entrepreneurs.

RCA changed the culture of the television industry to gain acceptance of all-electronic color television broadcasting. And it went on from there to change the entire culture of American home entertainment. With 90 percent of all television sets now able to receive color broadcasts, television is in a position to compete with film and other entertainment media in quality of visual image. This, in turn, has opened markets for hundreds of other products, including VCRs and satellite dishes. RCA's innovation shifted

American culture in a direction favorable to RCA's products and services in television.

The lesson is that innovation can and often does create, change, or even destroy an entire market. Indeed, when I encouraged you earlier in the book to "ask the innovative question," I may have understated the scope of that question. Managers must think boldly. Often, the right question is not "how can we make this better?" but "how can we change the habits and desires of millions of people in a way that uses our technology to increase our profits?"

A "people don't do that" attitude is the enemy of innovation. "People didn't" get money out of machines before Diebold introduced the ATM, but the customers were happy to change their habits. "People didn't" spend $15 to send a piece of paper across the country in a single day, but Federal Express and its competitors were not afraid to create an entire new market in overnight delivery. Companies like Diebold and Federal Express changed customer needs. They literally created an industry that hadn't existed before. In the future, every successful manager will have to know how to innovate the marketplace.

Of course, customers cannot be changed wholesale overnight, a lesson Walt Disney learned with the demise of his City of Tomorrow project. Disney did not design his city to pander to his customers' fantasies; he set down rules the customers had to follow. But culture doesn't evolve that way. Instead, targeted market segments can be changed with determined effort and directed signals. The initial changes can then be managed so that they build an acceptance of the innovation throughout society, so long as the innovation is basically sound in concept, design, and economics. The success of the credit card presupposed cultural acceptance of a "buy now, pay later" philosophy even though American culture had always equated debt with sin. Once new economic assumptions had been accepted, the credit card could be adopted also.

Personal computers required even greater cultural change to gain acceptance. But computer firms fostered that change through advertising and public relations campaigns geared not only to brand name recognition but also to creating an awareness of personal computers as a dominant cultural force. Apple, followed by other manufacturers, established a program to give away computers to schools in order to educate customers. This is just one example of how a company can successfully embed a new product into cultural expectations.

In conclusion, there is no end to opportunities for innovation. We have only scratched the surface in coping with change and managing innovation. Opportunities abound throughout all companies, in all industries.

Most of the company experiences herein illustrate product innovations. Manufacturing cases were fewer in number. While I have had many experiences in applying advanced technology to manufacturing, this area has not progressed as rapidly in the United States as have product innovations. Many of the concepts discussed in this book are equally effective for manufacturing, and it is in this sector where the next major thrust for innovation in the United States is most likely to occur.

Until now, manufacturing innovations have not progressed rapidly because of traditional, entrenched cultures. Yet as these cultures change, encompassing new organizational concepts in dealing with the blue- and white-collar work force, America's smokestack industries will prove to hold enormous innovative potential. As the microprocessor extends the life cycle of many of these industries' products and even create market opportunities through new products, so will electronic and computer systems change the manufacturing processes for these products.

This book is a hopeful one. Its stories illustrate that companies can change. In the right environment internal development programs can create opportunities so that even lagging companies can become innovative leaders. It is my hope that this book has provided insights and challenged managers at all levels. With these managers rests the greatest potential for innovating change.

ABOUT THE AUTHOR

John S. Rydz is the corporate vice president of technology for the Emhart Corporation, a $2 billion diversified global corporation. At Emhart he develops worldwide technical strategies, fosters emerging technology for Emhart's products and processes, and has developed programs for improving the company's innovative position in its products and manufacturing processes.

Prior to joining Emhart, Rydz was the vice president and chief technical officer for the Sewing Products Group of the Singer Company, where he was responsible for the worldwide technical activities of Singer's Sewing Products Group. While at Singer he launched and directed the Singer Athena 2000, the world's first electronic sewing machine.

Rydz was also a corporate vice president at Diebold Incorporated, where he was responsible for Diebold's initial development of ATMs. He was a research manager at Addressograph-Multigraph, where he directed AM technical activities into plastic credit card systems; the executive vice president and a director of Nuclear Corporation of America; and a staff member of the senior executive vice president for the Radio Corporation of America.

He is a graduate of the Massachusetts Institute of Technology with a B.S. in Physics. He received his M.S. in Physics from the University of Pennsylvania and has studied business at Case Western Reserve. Rydz is a member of

MIT's Industrial Advisory Council on Manufacturing and Productivity, the University of Connecticut Materials Technology Council, the University of Hartford Engineering Council, the IRI's Management Training Council, and the Greater Hartford Chamber of Commerce Technology Board Council.

INDEX